FOOD SCIENCE AND TECHNOLOGY

OLIVE OIL

SENSORY CHARACTERISTICS, COMPOSITION AND IMPORTANCE IN HUMAN HEALTH

FOOD SCIENCE AND TECHNOLOGY

Additional books in this series can be found on Nova's website under the Series tab.

Additional e-books in this series can be found on Nova's website under the eBooks tab.

FOOD SCIENCE AND TECHNOLOGY

OLIVE OIL

SENSORY CHARACTERISTICS, COMPOSITION AND IMPORTANCE IN HUMAN HEALTH

THORE FRITJOF

AND

BERTIL HENNING

EDITORS

nova biomedical

New York

NOTICE TO THE READER

The Publisher has taken reasonable care in the preparation of this book, but makes no expressed or implied warranty of any kind and assumes no responsibility for any errors or omissions. No liability is assumed for incidental or consequential damages in connection with or arising out of information contained in this book. The Publisher shall not be liable for any special, consequential, or exemplary damages resulting, in whole or in part, from the readers' use of, or reliance upon, this material. Any parts of this book based on government reports are so indicated and copyright is claimed for those parts to the extent applicable to compilations of such works.

Independent verification should be sought for any data, advice or recommendations contained in this book. In addition, no responsibility is assumed by the publisher for any injury and/or damage to persons or property arising from any methods, products, instructions, ideas or otherwise contained in this publication.

This publication is designed to provide accurate and authoritative information with regard to the subject matter covered herein. It is sold with the clear understanding that the Publisher is not engaged in rendering legal or any other professional services. If legal or any other expert assistance is required, the services of a competent person should be sought. FROM A DECLARATION OF PARTICIPANTS JOINTLY ADOPTED BY A COMMITTEE OF THE AMERICAN BAR ASSOCIATION AND A COMMITTEE OF PUBLISHERS.

Additional color graphics may be available in the e-book version of this book.

Library of Congress Cataloging-in-Publication Data

LCCN: 2017955342
ISBN: 978-1-53612-563-4

Published by Nova Science Publishers, Inc. † New York

CONTENTS

PREFACE

In Chapter One, D.M. Salazar, I. López-Cortés, and D.C. Salazar-García give an overview of the composition of fatty acids, tocopherols, and polyphenols in oils from the main five olive areas. They conclude that it is advisable to consume extra virgin olive oils during the prenatal period, childhood, and in combating disease. In Chapter Two, Hédia Manai Djebali and Imen Oueslati examine the link between olive oil and human health, finding that the amounts of sterols depend on olive variety, ripening degree, oil extraction procedures, agronomic, and climatic conditions. In Chapter Three, Malika Drira, Ons Rekik, and Mohamed Bouaziz review the composition of olive oil, determining that it is rich in unsaturated fatty acids and also identifying 20 phenolic compounds from a plethora of phenolic types. María Dolores Alvarez, Wenceslao Canet, María José Jiméne, and Cristina Fernández present a study where samples of mashed potatoes enriched with extra virgin olive oil were perceived as softer in Chapter Four. In chapter five, Imen Oueslati and Hédia Manai explore the characteristics of the volatile compounds, as well as the enzymes involved, in different Mediterranean olive oil categories.

Chapter 1 - The production of Extra Virgin Olive Oil (EVOO) in Spain is very high, it reached 1 million tonnes in the last olive oil campaign, with over two million hectares planted with olive trees. This crop is distributed in over six different bioclimatic zones and with more than 100 cultivars, many of them native from a pomological point of view. Among the olive areas of Spain, Andalusia, Extremadura, Catalonia and Valencia stand out, next to the Central Region (Castilla-La Mancha). Each one of them is respectively dominated by Picual, Manzanilla, Arbequina and Farga cultivars alongside Cornicabra, all highly diversified. In response to this diversification of olive groves in Spain, the authors have studied the composition of the majority of oils of the main five olive areas, both in their saponifiable composition in fatty acids and their unsaponifiable fraction, tocopherols and polyphenols among them. The characteristics of many of these oils with a high content of oleic acid and antioxidants, make them very suitable as nutrients and healthy products. They also prevent cardiovascular diseases, cancer, and especially diabetes and other diseases that show a significant increase among the population, and more specifically, in child population. EVOO consumption is considered to be very suitable both in the prenatal period and in childhood, and there are Spanish oils that are especially convenient for these life stages because of their balance, softness and very pleasant aromas.

The current use of EVOOs in food is a very suitable requirement in the face of an increase in nutrition-related diseases, since we really are what we eat and how we eat it, therefore its use is advisable both in hospital diets and catering, and domestically.

Chapter 2 - The chemical composition of olive oil has been traditionally clustered into major and minor compounds; the formers are, in a large part, responsible for the olive oil main characteristics while the latters are markers for their peculiarities. The minor fraction also known as unsaponifiable fraction is rich in minor metabolites such as sterols, triterpene dialcohols, phenolic compounds, tocopherols, hydrocarbons, pigments, etc. Phytosterols make up the greatest proportion of the unsaponificable fraction and are the most used for authenticity and discrimination purposes. During the last years, sterols have received particular attention because of their capability to lower serum cholesterol levels in the human body resulting in significant reduction in the risk of cardiovascular disease. Furthermore, they are considered as anti-inflammatory, anti-bacterial, antioxidant and anti-cancer agents. The analysis of the olive oil sterol contents is usually carried out according to the official method provided by the EU and the IOC. Several authors highlighted that the amount and the composition of sterols depend on many factors such as olive variety, ripening degree, oil extraction procedures, agronomic and climatic conditions.

Chapter 3 - Different monovarietal extra virgin olive oil cultivars in Tunisia (Chemlali, Chetoui, Oueslati, Neb Jmal, Toffehi, Fakharri, Gemri, Zalmati, Gemri-Dhokkar, Chemchali and Dhokkar) having different chemical composition, were investigated in terms of their quality indices, sterol content, phenolic composition and volatile composition, oxidative stability and sensory profile to show their classification.

The identification of phenolic compounds was based on separation by high performance liquid chromatography (HPLC) equipped with Diode Array Detector (DAD) followed by chromatography-mass spectrometry analysis (LC-MS/MS). However, the qualitative and quantitative differences in phenolic composition were observed among cultivars, thus identifying twenty phenolic compounds belonging to different phenolic types. Indeed, the sensory evaluation remains the gold standard for assessing the aroma. Olive oils are the only foods where sensory evaluation is a legal requirement. Oil samples are scored and classified according to three positive attributes: Fruity, Bitter and Pungent. The antioxidant activity of olive oil extracts was evaluated by DPPH. Olive oil is rich in antioxidants (like vitamin E and poly phenols). Moreover, it is rich in mono unsaturated fatty acids unlike saturated fatty acids. The results indicate that the cultivar influence the quali -quantitative production of volatiles and consequently the final aroma of the oil.

Chapter 4 - Growing awareness of the link between diet and health is fast changing consumer habits, so that there has been increasing demand for foods with health-enhancing properties. Extra virgin olive oil (EVOO) has important nutritional characteristics linked to its biophenol content and has very important antioxidant properties. The aim of this chapter was to study the effect of the addition of EVOO on instrumental textural properties, sensory texture profile analysis (TPA), and microstructure of fresh and frozen/thawed mashed potatoes formulated without and with added cryoprotectants (kappa-carrageenan [κ-C] and xanthan gum [XG]). EVOO behaves as soft filler due to droplet aggregates, whereas addition of cryoprotectants led to more structured mashed potatoes (MP) thanks to the gelling properties of κ-C. Both the percentage of added EVOO and processing had a much less significant effect on the texture of the MP containing κ -C and XG, evidencing the ability of this biopolymer blend to impart freeze/thaw stability. All samples with added EVOO were perceived as significantly softer and creamier than the samples without EVOO, whereas all MP samples with added cryoprotectants were perceived as significantly thicker and creamier than those without hydrocolloids. The results have shown that although instrumental textural data were able to explain differences in consistency perceived, structural information is needed to understand differences in creaminess. Back extrusion test is recommended to industry as practical quality control tool in the commercial production of MP with added EVOO.

Chapter 5 - Virgin olive oils (VOOs) represent the main source of fats in the countries of the Mediterranean basin where the production of olive oil is concentrated. Its quality is based upon consumer perceptions of aroma, and the absence of sensory defects. Volatile compounds responsible for the olive oil aroma count approximately two hundred compounds belonging to several chemical classes, identified via gas chromatography-mass spectrometry. The main characteristics of the volatile compounds of the several Mediterranean olive oil categories and the several enzymes involved in the pathways and in their activity were investigated. The synergy and antagonism processes between volatile compounds contribute to the sensory evaluation of VOOs, so it is of great interest to ascertain the relationships between sensory attributes and the volatiles responsible for them. Aroma compounds responsible for flavours, and for off-flavours including factors affecting the volatile fraction, are carefully examined, also considering the causes that give rise to off-flavours. New approaches were adopted to the definition of EVOO quality that might include also healthy and sensory parameters and on the agronomic and technological factors that most influence their variability.

In: Olive Oil ISBN: 978-1-53612-563-4
Editors: T. Fritjof and B. Henning

Chapter 1

OLIVE OIL: COMPOSITION AND HEALTH BENEFITS

D. M. Salazar[1,*], I. López-Cortés[1] and D. C. Salazar-García[2]
[1]Departament of Vegetal Production,
Universitat Politècnica of València, Valencia, Spain
[2]Grupo de Investigación en Prehistoria IT-622-13
(UPV-EHU)/IKERBASQUE-Basque Foundation for Science
Vitoria, Spain

ABSTRACT

The production of Extra Virgin Olive Oil (EVOO) in Spain is very high, it reached 1 million tonnes in the last olive oil campaign, with over two million hectares planted with olive trees. This crop is distributed in over six different bioclimatic zones and with more than 100 cultivars, many of them native from a pomological point of view.

Among the olive areas of Spain, Andalusia, Extremadura, Catalonia and Valencia stand out, next to the Central Region (Castilla-La Mancha).

* Corresponding Author Email: dsalazar@upv.es.

Each one of them is respectively dominated by Picual, Manzanilla, Arbequina and Farga cultivars alongside Cornicabra, all highly diversified.

In response to this diversification of olive groves in Spain, we have studied the composition of the majority of oils of the main five olive areas, both in their saponifiable composition in fatty acids and their unsaponifiable fraction, tocopherols and polyphenols among them.

The characteristics of many of these oils with a high content of oleic acid and antioxidants, make them very suitable as nutrients and healthy products. They also prevent cardiovascular diseases, cancer, and especially diabetes and other diseases that show a significant increase among the population, and more specifically, in child population.

EVOO consumption is considered to be very suitable both in the prenatal period and in childhood, and there are Spanish oils that are especially convenient for these life stages because of their balance, softness and very pleasant aromas.

The current use of EVOOs in food is a very suitable requirement in the face of an increase in nutrition-related diseases, since we really are what we eat and how we eat it, therefore its use is advisable both in hospital diets and catering, and domestically.

Keywords: food, EVOO, health, composition

INTRODUCTION

There is evidence of the consumption of fruits and its liquid extractions in diets already in the Copper Age (6th millennium BC). In the 7th millennium BC, olive oil began to be used as a cosmetic and for basic medical purposes. In the 9th and 8th century BC oil was used for food and also to take care of burnt skin, dermatitis, stomach and intestine problems, as well as sun protection [1].

Hippocrates (460-377 BC) used olive oil as medicine both for burns and all kinds of inflammations and digestive problems.

Pliny (24-71 AB) in his “Historia Naturalis” made a list of 48 medicines made from olive oil [2].

Studies on oil and health are and have always been very important. If we read ancient treaties on pharmacology, medicine and agronomy, the role of oil in health is highlighted in many of them, but they also speak of the virtues of the olive tree, especially of its leaves and fruits, as did Dioscorides, citing its many beneficial actions and its role as a condiment-preservative in many foods [3].

Ancient and modern pharmacopoeias have included numerous preparations based on olive oil, considering its therapeutic use as a remedy for many maladies, with beneficial effects in many of the human organs and systems, and in all phases of life.

In the Spanish pharmacopoeia in a very special way, as well as in other European pharmacopoeias, a broad monograph is dedicated to the remedies that olive oil can provide, even recommending its use in injectables, with a specific role in gynecology, in childhood and old age, but also in youth and adult ages. It has always been considered at least a protector of health.

After speaking of the great longevity of the olive tree and the possibility of its use in marinades [3], Font-Quer describes several marinades and recommends them as food, he indicates that according to various ancient authors, olive tree leaves contain glycoside oleuropein, in a proportion close to 0.75% although according to Power and Tutin it is not a pure substance but a mixture of several bitter principles, including various resins and alcohols. This author mentions that "olive oil takes a long time to turn rancid and for this reason, it is very useful in pharmacy to prepare ointments and creams" as it is still done today, and specifies that the leaves are considered "antipyretic, they lower the blood, this is to reduce the tension of the circulatory system and cause urine," Quer also indicates that "the effects of olives are usually to stir appetite and fortify the stomach, they make it easier to evacuate the intestine, only overeating causes damages."

This same author mentions olive oil as food in the Mediterranean countries, an ingredient to prepare ointments for burns, for which he indicates several formulas that are regarded as very good healing agents.

When it comes to the virtues of olive oil, Font-Quer mentions it is a laxative, an emetic in certain preparations and therefore an antivenin, an emollient mixed with mauve which helps avoiding any kind of constipation.

The authors of A Practical Guide on Medicinal Plants [4] indicate that olive oil contains very beneficial oleic acid, phitosterina lecithin and enzymes, in addition to a bitter principle and pigments, and they recommend it as a tranquilizer, an anti-inflammatory, a laxative, a stimulant of the biliary secretion and a nutrient. Regarding its leaves, they indicate that they have a bitter principle, certain important oleuropein, flavonoids, organic acids, tannins, carotenoids and evidence of essential oils. Olive tree leaves are attributed a hypotensive, antispasmodic, vasodilator and an astringent effect.

The guide to Medicinal Plants and Flowers [5] does not speak of the properties of oil, but indicates that the bark is antipyretic and astringent. The leaves are considered to be one of the best vasodilator agents, they are diuretic, antipyretic and anti diarrheal. They are prescribed in cases of hypertension, arteriosclerosis and diabetes. They prevent chest angina and kidney damage, they facilitate the non-formation of bladder stones, and they are regarded as sources of active principles with a protective hepatic function.

Much has been discussed about the position and importance of oil and in particular of EVOO (Extra Virgin Olive Oil) in the Mediterranean diet [6, 7, 8] and in food pyramids, at least about the recognition of this diet as appropriate [9, 10]. It reduces deaths due to cardiovascular diseases [11] and has shown excellent results in inflammatory processes treatments [12].

It is important to establish the level of polyphenols and monounsaturated fatty acids as antioxidants and anti-inflammatories [13] along with their nutritional properties and their ability to maintain the oxidative stability of oils for longer. Also their interesting content in carotenoids and chlorophyll [14].

Many other authors have boosted the search and development of new analytical techniques by HPLC-DAD of hydroalcoholic extracts that coupled to a mass spectrometer improve and facilitate the study of these compounds, and especially of tyrosol and hydroxytirosol [15], in order to know their content in different oils either from different cultivars or the same cultivar in different regions.

Currently, it is very important to know the composition of varietal oils in detail [16] and to know what is the concrete role played by their most noble components in health, the role of the predominant monounsaturated fatty acid (oleic acid) [17] and its proportion in relation to the saturated and polyunsaturated fatty acids, the role of polyphenols, tocopherols and phytosterols. Extra virgin olive oil (EVOO) also contains other components whose biological properties cannot be forgotten, such as chlorophyll, which plays a stimulating part in many organs and accelerates the healing process, vitamin A and magnesium, both with a possible anti-cancer action, etc.

For a long time, food was one of the main control resources for many diseases. In the Mediterranean culture, olive oil has always been an important resource, it remains so to this day and therefore its therapeutic function on diabetes mellitus, systolic pressure decrease and as a diastolic pressure moderator, in the prevention of cardiovascular diseases, etc.

The regular consumption of EVOO is considered suitable [18] for mothers in prenatal periods, children, adolescents and especially the elderly and in hospital diets [19].

The role of olive oil in preventive medicine is presently expanding, and it is always associated with healthy habits of behavior. For this reason, together with food and activities, olive oil plays an important role in the pyramid of the Mediterranean diet, which is nowadays regarded as a powerful support in the prevention of cancer [8, 20]. Interest in olive oil has reemerged [21], especially EVOO in its own right [22] and there is also a greater interest in its specific compounds [23], especially monounsaturated fatty acids, polyphenols, tocopherols [24], etc. It is still very important to continue to deepen in their specific role and that of other components not only as antioxidants [25, 26], but also in the control of chronic diseases such as arteriosclerosis, heart diseases [27], Parkinson's disease [28], the regulation of cholesterol plasma level, obesity, and hypertension control [29], etc. Therefore, the active component characterization of monovarietal oils is essential [30, 31, 32].

At the moment, there is a clear need to replace the current industrial pastries made with different types of oils and fats, with EVOO products and the daily consumption of between 5 and 10 fruits or vegetables.

Olive oil, in addition to being rich in polyphenols and other antioxidant compounds [33, 34, 35, 36], has a superb balance between monounsaturated and polyunsaturated fatty acids. It is rich in the minor components related to the main current pathologies and it has a cellular anti-proliferation effect [37].

EVOO's antioxidant activity is attributable to henolic compounds, tyrosol and hidroxytirosol, lignans, flavonoids, carotenoids and chlorophyll and their derivatives, and squalene, a triterpene present in olive oils [22].

The high content in monounsaturated and in particular in oleic acid is now considered to be, without a doubt, very positive from a healthy, dietary perspective. In the studies carried out in the west of the Mediterranean, the following oils contain more than 80% of oleic acid: Arbequina, Carrasqueña, Cornicabra, Del Pomet, Hojiblanca, Llumero, Picudo, Seniero, Serrana and some Villalonga [38].

In nutrition, lipids are essential foods provided they do not exceed 30% of the total energy intake and this intake should be consistent with age, weight, and possible sensitivities to pathologies of the consumer, thus it is recommended that out of this total amount, between 12 and 20 per cent should be monounsaturated fats.

The general functions of fats, as well as being part of the cell membrane, are metabolic functions, they also facilitate the absorption of fat-soluble compounds, the synthesis of several hormones and secretions, etc.

The importance of EVOO in health is increasingly recognized, which is why a lot has been studied on polyphenols [39] oleuropein [40], hydroxytyrosol [41], all capable of partially inhibiting platelet aggregation [42, 43] and other compounds that are also present in the leaf of the olive tree [44, 45, 46] by insisting on their bactericidal properties [47], especially in cases of colitis and diarrhea, due to the high content of polyphenols in its leaves [48] which show an important antioxidant activity [49]. Nowadays, the extraction of polyphenols with an anticancer effect [44] and the oleanoic acid [50] of the leaves has become habitual, since they are considered to be a potential treatment for various cancers [35, 51, 52].

We shouldn't forget the oleuropein olive content [53], it is really an excellent antioxidant and an ancient food, an antioxidant and antimicrobial [54], although the excessive salt content of some preparations can lead to hypertensive disorders.

Monounsaturated fatty acids are particularly interesting, olive oil is rich in two of them: oleic acid (cis-9 octadecanoic) and palmitoleic acid (9cis-hexadecanoic) which are always beneficial to health and especially in people who have diabetes or other metabolic syndromes. These fatty acids act by decreasing the total cholesterol and LDL cholesterol when they replace saturated fatty acids, reducing triglycerides and increasing or maintaining the level of HDL cholesterol. They are also antihypertensive, antithrombotic, anti-inflammatory and contribute to the improvement of the endothelial functions.

Bioactive EVOO and VOO Components and Health Benefits

The high content in monounsaturated fatty acids (MUFAs), especially oleic acid, which plays very clear role in nutrition, and the richness in polyphenols, tocopherols and phytosterols, which are important antioxidants, have risen the already very high interest on olive oil as an essential component and a symbol of the Mediterranean diet [55]. The nutraceutical industry is working with extra virgin olive oil (EVOO) [56] because it acts as a cardiovascular and cerebral-vascular protector, it prevents neurodegenerative disorders (Parkinson's Disease and Alzheimer's Disease), it is an anti-inflammatory and it plays an important role in some types of cancer. It also intervenes in certain syndromes of the human metabolism and is a general detoxifier. All these characteristics make EVOO an essential food component for health.

Monounsaturated fatty acids, of which olive oil is the richest fat, are now considered to be valuable compounds in the prevention of cardiovascular pathologies and various types of cancer [57, 58]. They are also important in the maintenance of a normal cholesterol level in blood, which has made many nutraceutical industries decide to use olive oil or some of its components in their products [59] and thus they study different methods of extraction [60] of these fatty acids, particularly polyphenols, as antioxidants, in different fruits of the forest as well as in some nuts, even in other organs of the olive tree and other species that are rich in these components.

Oleic acid is an oil that is associated with the ability to reduce the effects of oxidative stress in the human body, the control of the levels of LDL-cholesterol without altering HDL-cholesterol and it therefore reduces the risk of heart attacks and other cardiac pathologies.

This acid contributes to a better control of the formation of foamy cells and the control of the accumulation of macrophages in the circulatory system, it minimizes the formation of atheromas [61, 62], which explains its role in the reduction of heart problems, along with other minority components in EVOO and VOO. The proportion of monounsaturated, polyunsaturated and saturated fatty acids consumed in our diet is also involved in the prevention of death or complications from cardiac problems.

Currently, the high oleic acid content of an oil is associated with its ability to act as an antitumour agent [57], and it is specifically attributed an important role in the decline of the possibility of colorectal or prostate cancer.

It is necessary to consider that the content of oleic acid and these relationships have a mainly varietal genetic component, but are influenced by weather conditions and the state of maturation of the olive in the harvest stage. The origin of olive oil and its oleic acid content as well as the composition of the oil of wild olive trees [66] have been studied [63, 64, 65] in different Mediterranean countries.

Polyunsaturated omega-3 fatty acids from fish have been considered to be very significant for some time now, due to their antilipidemic, antiarrhythmic and anti-inflammatory effects, they decrease the incidence of coronary events and cardiac diseases. Without losing its importance, this fatty acid has given way to the great benefits provided by monounsaturated omega-9 acid (oleic acid) major in olive oil, along with omega-6, as parts of olive oil and have been given a lot of publicity in the food sector. A number of authors [67, 68] have recently stated the need to deepen our knowledge on the role of the different fatty acids present in EVOO and VOO and the relationships between them [69] in order to clarify in the deepest possible way, their role as dietary supplements that are favorable to the reduction of cardiovascular pathologies [69] and in the treatment of various pathologies of the nervous system and even multiple sclerosis [70].

Some time ago, it was stated that [71] monounsaturated fatty acids contribute to the delay of cognitive faculty loss in older people as they avoid, among other things, neuronal damage produced by free radicals [72].

The fatty acid composition of a newly obtained oil that meets the specifications to be considered an EVOO (IOOC, 2013) clearly depends on the cultivar at issue, its genetics, but also the state of maturation of the olive and the ecological-geographic conditions in which it is cultivated [73, 74, 75, 76]. In the data reported in Figure 1, we always sought the same state of colorimetric maturation [77, 78, 79, 80], although collection has always been early, it has not been simultaneous [81]. This collection has always been carried out in experimental plots and traditional growings located next to the West of the Mediterranean, in the Levantine area of the Iberian Peninsula, although some cultivars do not originate nor are common in commercial plantations in these areas. The analyses have been carried out according to the rules and techniques indicated by AOCS (1998), CODEX and [82], evidently [14] the quantitative values of the components may slightly vary according to the area of origin of the EVOO. The oleic acid content, just like other components, may slightly decrease or increase depending on the cultivar [76] the data shows the average values of 16 olive trees for five olive oil year campaigns.

Table 1. Monounsaturated fatty acid composition and percentages of SFAs, PUFAs, MUFAs in the main oils studied

Variety	C18:1 (oleic ac.) %	C16:1 (palmitoleic ac.) %	ΣSFAs %	ΣPUFAs %	ΣMUFAs %
Aguilar	79,21 ± 0,88	1,42 ± 0,02	16,29	4,56	80,63
Blanqueta Enguera	61,95 ± 0,74	1,52 ± 0,06	21,35	17,51	63,47
Borriolenca	77,89 ± 0,51	0,77 ± 0,02	13,52	6,98	78,66
Cabaret	76,77 ± 0,91	1,95 ± 0,02	13,35	10,23	78,72
Callosina	76,06 ± 1,85	1,18 ± 0,34	17,42	9,12	77,27
Carrasqueña	83,98 ± 0,80	0,52 ± 0,02	11,52	4,46	84,51
Cuquillo	75,87 ± 0,54	1,31 ± 0,01	16,22	6,73	77,19
Changlot Real	77,06 ± 1,77	0,51 ± 0,05	14,36	7,45	77,57
Del Patró	72,60 ± 0,27	1,87 ± 0,04	15,77	9,94	74,47
Del Pomet	79,45 ± 0,36	1,24 ± 0,02	14,81	3,98	80,69
Figuereta	79,38 ± 0,58	0,42 ± 0,01	10,91	9,61	79,81
Genovesa	77,53 ± 2,34	0,62 ± 0,15	14,53	7,24	78,15
Gileta	69,68 ± 0,96	1,14 ± 0,04	12,98	13,49	70,82
Grossal	71,75 ± 0,05	0,58 ± 0,01	15,64	11,79	72,33
Lloma	83,84 ± 0,26	0,40 ± 0,03	10,66	5,64	84,24
Llumero	80,52 ± 1,24	0,60 ± 0,20	12,91	6,42	81,12
Marfil	78,94 ± 0,75	0,59 ± 0,01	13,23	6,89	79,53
Mas Blanc	74,79 ± 0,76	1,19 ± 0,02	12,65	11,92	75,99
Millareja	70,57 ± 0,48	0,97 ±0,05	16,45	12,22	71,54

Variety	C18:1 (oleic ac.) %	C16:1 (palmitoleic ac.) %	ΣSFAs %	ΣPUFAs %	ΣMUFAs %
Monteaguda	71,53 ± 1,04	1,23 ± 0,01	19,74	7,66	72,76
Morons	73,75 ± 1,49	0,65 ± 0,02	11,95	14,19	74,40
Morruda	72,81 ± 1,34	0,85 ± 0,02	15,76	10,9	73,66
Negra	73,32 ± 1,10	0,46 ± 0,01	15,23	11,57	73,79
Piñonera	69,15 ± 0,71	1,49 ± 0,02	17,72	11,72	70,64
Romana	68,53 ± 0,34	2,04 ± 0,02	20,68	8,82	70,58
Rotja	66,84 ± 0,90	1,08 ± 0,02	17,50	9,85	67,93
Rufina	76,82 ± 0,19	1,15 ± 0,01	16,22	6,07	77,97
Seniero	82,76 ± 0,44	0,76 ± 0,01	11,02	6,39	83,53
Valentins	82,29 ± 0,46	0,716 ± 0,01	11,28	5,00	83,00
Vallesa	73,60 ± 0,11	0,45 ± 0,07	14,35	11,76	74,05
Vera	74,19 ± 0,75	0,97 ± 0,02	15,54	6,57	75,16

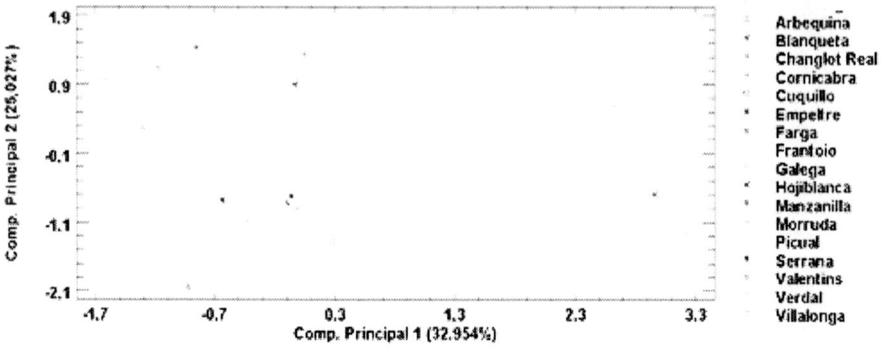

Figure 1. Tocopherol and polyphenol PCA in the main varieties.

On the other hand, different authors indicate that the geographical, agronomic and technological conditions influence the quality and sometimes the composition of the oil [64], especially in their phenolic compounds [65].

The diversity among EVOO is therefore important and it is related to the specific campaign to which they belong, due to the influence of the different temperatures and rainfall or irrigation. By using MRI we can differentiate the area of origin of the oils [74] and secure the area of cultivation which in our case is homogeneous. The techniques of cultivation can also influence the composition of the oils [75], which are also the same, or very similar and always in conditions of low water availability in counties with a rainfall average between 300 and 420 mm per year.

One of the most interesting components for health of EVOO and VOO are phenolic compounds, whose activity was controversial for a period of time [83, 84], but after being revised [13] it was concluded their great importance and, in particular, the fact that they are bearers of ortho diphenolic groups [85].

Nowadays, polyphenol analysis techniques [15, 86] have been greatly developed, as well as the valuation of its antioxidant properties [85]. Many authors refer to this very important effect of polyphenols in olive oil [87], since they are considered to be efficient in the treatment and prevention of the inflammations [88] involved in numerous diseases and that in different formulations have applications as nutraceuticals (pharmaceutical accessories), cosmetics, and in the food industry. At the Paris International Conference on Polyphenols (2015) it was stated that "economically feasible approaches are still needed for the exploitation of biophenols by products arising from processing of plant derived raw materials." Polyphenols also have clear antioxidant properties [89, 90, 91], since they block the oxidation process in its initiation stage, they oxidize themselves forming peroxide compounds of a more stable nature or act on the radicals already formed by avoiding the cascade formation of new radicals (CODEX SJAN 33-1981 of 1989). They also take active part in the prevention of certain types of cancer [92, 93] (including prostate and colorectal cancer [94] and colorectal [95]) in cardiovascular diseases, since they are considered vasodilators [96] and antiaterogens, specifically inhibitors of leukocyte-5-lipoxygenase [97]. All of them play an important role as neuroprotectives [98].

Polyphenols are antioxidants due to their role as free radicals and they are considered to be anti-inflammatory [99], antihepatotoxic, antidiarrheal, anti-ulcerous in the digestive system [100], anti-allergic, anthelmintic and also have certain antiviral effects. Furthermore, polyphenols are highly valuable for their protective effects from cardiovascular diseases, an important effect capable of reducing neurodegenerative processes [18], these great expectations have helped achieve a greater progress in the evaluation of the antioxidant power of these compounds.

They are also considered preventive of cerebral ischemia problems and others related to neurotransmission [101] such as Alzheimer's disease [28], Huntington's disease [102], Parkinson´s disease, certain peripheral neuropathologies, problems in the spinal nervous system and even multiple sclerosis.

In the wild olive tree [33] we mentioned contents between 23.6-25 and 81-92.4 mg/kg, in the oils studied in the west of the Mediterranean [31] we have obtained total values between 462.81 and 126.83 ± 0.94 ± 3.18.

We must remember that all EVOOs and VOOs for commercial use do not have the same polyphenol content [103, 104] and we understand that this content should be specified on labels today.

The most important phenolic compounds in olive oil and leaves of the olive tree are polyphenols, especially oleuropein, which is an heteroside of elenoic acid, and hydrocortisols. They are both basically very powerful chelating agents of metal ions and abductors of free radicals[105].

Anyway, we must remember [106] that the place of origin of oil, the agronomic management of the olive trees and especially the extraction technologies, will greatly influence the total polyphenol content of the extra virgin olive oil [64].

Currently, talking about the global role of the components of EVOO is important, but we should focus even more on the specific role of each of its components like hydrocortisol, oleuropein, fatty acids, especially monounsaturated, phytosterols, tocopherols, almost all of them play a proven antioxidant role.

Hydrocortisol shows a high antioxidant activity [107] and it is considered to be very important because it prevents various neurodegenerative problemsa and it protects against brain synapses [108] [109, 110]. In addition, many authors [111, 112, 113] studied the mechanisms that interlink hydrocortisol to the activity of monocytes and their action on inflammation [114] and especially its role as a superoxides blocker [115] and its importance in adenylate cyclase control.

The effect of hydrocortisol at an endoplasmic reticulum level has now been confirmed in the attenuation of hepatocarcinoma [116] and the preparation of new drugs for cancer control and prevention [117].

Oleuropein is a sericoid with clear pharmacological properties [118]. It is an antibacterial, an antioxidant, an abductor and inhibitor in the formation of free radicals [90], it is anti-inflammatory, it has shown an effect in obesity control, it is antiaterogenic and therefore plays a cardioprotective and regulating role on hypertension [119]. It is even antiviral [87].

Oleuropein of the leaves of the olive tree and olive oil has more recently been tested as a cell anti-proliferation agent [120, 121], also in breast and ovarian tumors with an inhibitory action of their metastases [122], with action on certain kinases and the mitochondria of cells in rapid growth [123].

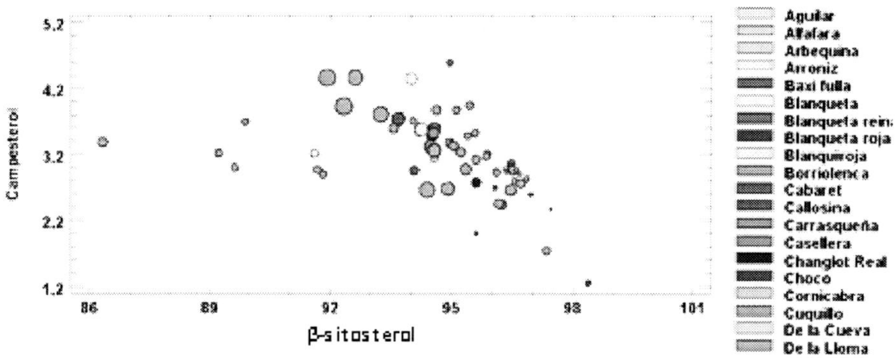

Figure 2. Majoritarian composition of the three major sterols according to cultivars.

On oleic acid, whose content varies according to the varietal origin of the olive oil, and which is also known as Omega 9, among other basic features, we know that it reduces the risk of atherosclerosis and thrombosis.

Polyunsaturated fatty acids generated great excitement in the past due to their beneficial health effects, especially when they were shown to decrease the risk of atherosclerosis. Nowadays, although they are still considered important, at least series omega 6 and omega 3, it has been proven [22] that monounsaturated fatty acids have the same effect and they are more complete.

We should mention as fatty acids in olive oil of the series Omega 3, linolenic acid (6C9C12C) with 18 carbon atoms and 3 double links and arachidonic acid (5C8C11C14C eicosatetraenoic) with 20 carbon atoms and 4 double links. The most important in the series Omega 6, is linoleic acid (estearidonic 6C9C12C15C octadecatetranoic) with 18 carbons with 4 double links which, among other things, reduce the risk of atherosclerosis.

Olive oil contains linoleic acid, which is defined as an essential fatty acid for the human being and the lack of which today is associated with skin problems (scaly skin, eczema and abnormally pale skin), imbalances in prostaglandins, poor control of inflammation and blood pressure associated with thrombotic and vasoconstriction problems, especially in the peripheral circulatory system. On the other hand, linoleic acid is key in the control and metabolic regulation of diabetes, so the high content of linoleic acid in olive oil compared to other oils makes it appropriate to consider EVOO as a nutrient whose limited but constant consumption should be regarded as a true pillar in an adequate nutrition for health maintenance.

Phytosterols are [124, 125] components that positively support health, but certain authors indicate that the stigmasterol content must always be well below the content in campesterol.

The relative values between these compounds are the basis for the detection of mixtures of EVOO and seed oils, which allows to clearly differentiate between oil as a natural juice of the olive and other extraction oils, differentiating also refined olive from the VOO and EVOO, though their content in these oils very much depends on the cultivar [126].

There are more than 200 types of steroils in different vegetables. The main sterols in olive oil are β-sitosterol and campesterol, and stigmasterol to a much lesser extent. All of them play outstanding roles in maintaining human health. This is due to their hypocholesterolemic effect, they reduce both total cholesterol and LDL cholesterol, acting not only in the intestinal absorption of cholesterol, but also in its synthesis in eliminating systems.

In this regard, we find particularly interesting the role of sterols as an anti-inflammatory and their anticancer effects, with their action on the immune system [127].

Phytosterols [124], both from EVOO and other products, have significant effects on health. They are antibacterial, antifungal, they prevent the appearance of ulcers in the digestive system, they have anti-inflammatory properties, they are antioxidants and therefore preventers and modulators of tumor growth, especially in breast cancer [128] and they play an interesting role in the minimization of degenerative processes associated with age, and in particular in the prevention of macular eye degeneration in people of advanced age.

Phytosterols and specifically different sterols and squalene are considered powerful bioactive substances present in olive oil, because they reduce intestinal absorption of cholesterol and therefore help maintain low levels of cholesterol in blood and clearly contribute to the prevention of heart attacks [129].

β-sitosterol, which is the majoritarian phytosterol in olive oil, reduces the absorption of compounds with nitrogen (nitrites, nitrosourea, etc) which induce digestive tumors and especially colon tumours [124], it also reduces the amount of cholesterol, especially LDL cholesterol by interfering with its absorption [130], reducing LDL cholesterol and not affecting the serum levels of HDL-cholesterol. On the other hand, β-sitosterol has for a while been considered suitable for the treatment of prostatic hyperplasia [55, 131] and breast cancer control [132]. In general, the increase in the consumption of phytosterols and the reduction of cholesterol, clearly reduce the incidence of different types of cancer. Campesterol is considered today a chemical health parameter of olive oil.

Tocopherols are relevant in the discrimination of the stable isotope analysis [133], they take a strong antioxidant action, α-tocopherol is the most active biologically speaking [134], which is the one studied in our chapters and reflected, by varieties, in the table above. Its contents are also considered. It very useful in the detection of possible tampering in oils and it is important in cancer prevention [135]. It must be borne in mind that temperatures in the process of oil extraction can significantly reduce the tocopherol content.

α-tocopherol has a clearly defined role as an antioxidant and a free radical, it has been very much biochemically studied in the integrity maintenance of cell membranes against chemical contaminants, tobacco, alcohol, excessive exposure to sun and heat, application of radiotherapy, etc.

We must remember that the presence of free radicals, in addition to damaging fatty acids in cell membranes that they deteriorate, forces the rupture of fatty acids that form new free radicals (superoxides and hydroxides) and can form very toxic aldehydes at a cellular level by acting on the double links of polyunsaturated fatty acids and lipoproteins, forming peroxidized and malon-dialdehyde fragments (MDA) that have been proven to widely increase (up to 20-25 times) in the acute phase of myocardial infarction and 20-25 hours after a stroke. This progressive cascade of peroxides is associated with certain types of infections, allergies and chronic inflammation. In addition, MAD is considered to be one of the compounds that cause most damage to DNA structures that constitute genes, it being the cause of numerous genomic alterations associated with cellular or even organic pathologies.

Cancer and other numerous diseases such as atherosclerosis, cataracts, neurovegetative diseases (Parkinson's and Alzheimer's) are currently associated with the oxidation of specific structures such as the eye lens and the neural formations.

So we can summarize [134] that tocopherols (provitamin E) which are very much present in EVOO, are involved in:

- The cardiovascular system, avoiding or mitigating thrombus formation. They reduce platelet activity and decrease the possibility of forming atheromas and thrombus by reducing the risk of heart attack, angina pectoris, strokes, but also avoiding cramps, especially in the legs, in people with venous return problems.
- They stimulate the reparative power of a physical injury, by encouraging the formation of elastin fibers and collagen in the conjunctive tissue.
- They promote the formation of cellular responses in specific infections and damages by extra organic invasions, and stimulate the cellular immune system and they possibly have an effect throughout the immune system.
- Tocopherols are essential for the proper functioning of the reproductive system and fertility, although in this sense its deficiency is rare, in chronic patients who lack fat assimilation, the role of this olive oil component is more easily assimilated, it is the case of its role in preventing vitamin E deficiencies, in patients with cystic fibrosis and Crohn's disease, in both children and the elderly who are more vulnerable to this deficiency. In these cases they show weakness and muscle atrophy, jaundice, they are also involved in coagulation processes because they can alter the adhesion of platelets.

- Tocopherols and several tocotrienols present in olive oil [136] clearly contribute to the protection of cells against oxidative stress.

Tocopherols, vitamin E precursors, can be found in significant quantities in green leaves and in the germ of cereals and olive oil, in this case, preferably in the form of α-tocopherol, which contains, according to various authors, between 125 and 200 g/kg [30], between 97 and 224 mg/kg and between 150 and 430 mg/kg [137]. In our studies, the content in α-tocopherol has varied between 86.84 ± 0.56 and 343.01 ± 6.98, depending especially on the variety it is obtained from and thus taking a chemotaxonomic high value in oils and an undoubted value in the election of the oil, always EVOO.

Terpenes are also a set of organic compounds derived from an isoprene compound with five carbon atoms (2methyl-1,3-butadiene) that are also called terpenoids when they are reorganized in their hydrocarbon structure. Within the existing terpenes in olive oil, we should basically mention tetraterpenos or carotenoids, vitamin A precursors, but also triterpene squalene, and certain diterpenes, Vitamin A or retinol formers, from base tocopherols of vitamin K. Olive oil is considered to be essential in food, among other things due to its contribution in these three vitamins.

The carotenoids contained in oil accumulate in many tissues, plasma lipoproteins are the mean of transport to the different organs for which they are essential, they are precursors of vitamin A and responsible for the yellowing in olive oils, in which they are present in their three isomers α, β, and γ, related by their antioxidant role in the protection of nucleic acids, therefore they possess antiatherogenic and anti cancer effects (efficient against certain types of cancer), they maintain normality in the ocular system, skin and mucous, strengthen the immune system and protect our body against cardiovascular diseases.

We must remember that the process of discoloration in the refining of oils, including olive oil, will eliminate a large part of these pigments, especially β-carotene, which is the most abundant in EVOO, and whose content typically varies between 0.03 mg and 0.5 mg/100 g depending on the altimetry of the crop, the forcing of productions by excessive addition of nitrogen, the ripening index, the management of the olive from harvest to the mill and obviously the system itself, speed of extraction and proper preservation of the oil.

The consumption of squalene [138] is considered to be an oxygenic radical receiver and therefore a protector of the lipid peroxidation attributed, at least in the skin, to an excessive exposure to ultraviolet and other ionizing radiation, so that its high content in EVOO, since the process of refining olive oil reduces the contents of this valuable compound by more than 70%, causes that in diets in which EVOO is used, the squalene content is multiplied by 10, compared to other diets without this type of oil. Squalene consumption within the Mediterranean diet, rich in EVOO, can reach 200-450 mg/day. This is considered very suitable as a complement and a statin booster, an active hypocholesteromiant principle, because it makes a very efficient contribution to lowering blood levels of cholesterol and triglycerides in blood [134].

The chlorophyll content in olive oil, as a non-oily component, clearly determines the color of the oils that are obviously lost in the refining process. We must not forget its role as a facilitator of cell growth, and its part in the stimulation of the formation of blood cells (hematopoiesis).

The chlorophyll content is highly dependent on the concrete variety of olive tree from which olive oil is obtained and mainly from its state of maturation. For this reason and for the higher content in aromas and balance of the different components of olive oil, and therefore of its organoleptic quality, some early olive collections are thus recommended nowadays in order to enhance the quality and suitability for health of EVOO.

We should mention cultivars Picual and Koroneiki [31] among the oils with a higher chlorophyll content, if they are harvested properly as well as some wild olive trees.

One of the most serious problems for human health are high blood pressure and other cardiovascular pathologies, we must understand that normal averages between systolic pressure BP >140 mm Hg and the diastolic BP is >90 mm Hg in hypertension control components like phytosterols (β-sitosterol), triterpenes and especially phenolic compounds are essential.

Arteriosclerosis is a serious problem present in human population, it is a disease related to ageing and the natural wear of the circulatory system, identified sometimes with age itself, what is true is that a man is as old the vessels of his circulatory system [139].

Atherosclerosis is a progressive process of functionality loss in the circulatory system as a result of the accumulation of platelets and monocytes in its walls, which results in the loss of elasticity and partial obstruction, very problematic when it happens in vital arteries. The different variants of arteriosclerosis are the main origin of various diseases associated with old age, for example, lens, heart and locomotor apparatus damages and the cause of numerous deaths.

This disease has very variable extension and locations, affecting the aorta, the peripheral circulatory system, the vessels of the brain, the coronary arteries, the irrigation of the pancreas and kidney, etc.

Some time ago, an escalation in the consumption of animal fats and some other fats of vegetable origin was associated with an increase of coronary thrombosis. In addition, a clear increment in the possibility of unexpected blood clotting, which was associated with an increase in the content of saturated fats in plasma, which stimulate the formation of thromboplastin, when they exceed certain levels, they promote blood clotting and predisposition to the formation of atheromas.

The major role in the regeneration of vascular endothelial cells is studied by numerous authors [96] who insist on the role of the abundant polyphenols in EVOO and the Mediterranean diet in general, in the prevention and treatment of atherosclerosis, and even in the evolutionary processes of different types of cancer.

The beneficial effects of EVOO in hypertension control, as a consequence of the endothelial activity of vessels in response to different stimuli and substances such as nitric oxide, have been proven both in hypertensive and in normotensive individuals. Moreover, this is a result of the interaction of polyphenols, hidroxytirosol and some fatty acids present in olive oil, noting a greater antihypertensive effect in EVOOs which are richer in polyphenols.

Knowledge of the adequacy of nutrient intake is still considered insufficient and it is stated that "between 20-25% of the total calories consumed should come from fats that include essential unsaturated fatty acids, at least 1-2% in relation to the total calories, especially in highly active people, children and adolescents, this is ensured by a regular consumption of olive oil" [140].

We must remember that neither olive oil nor its components are drugs but only, and this is already quite something, preventives and nutraceutics [56] that can clearly help improve the state of human health.

References

[1] Mazzini I. (2000). L´uso dell´olio d´oliva nella medicina del mondo antico. *Mediazinhist J.* 35. 105-126.

[2] Flandrin J. & Montanazi Eds. (1997). *Storia Dell´alimentazione. Laterza* Ed. Roma-Bari. Italy.

[3] Font-Quer, P. (1981). *Plantas medicinales. El Dioscórides renovado. [Medicinals plants. Renoved Dioscórides*]. Labor Ed. Barcelona. España. 1033 pp.

[4] William A. R. & Thomson D. M. (1989). *Las plantas medicinales. [The medicinal plants.*] Blume Ed. Barcelona Spain. 220 pp.

[5] Poletti A. (1995). *Plantas y flores medicinales (I). [Flowers and medicinals plants (I)]*. Parramón Ed. Barcelona. Spain. 208 pp.

[6] Keys A. (1970). Coronary heart disease in seven countries. *Circulation*. 41 (1). 1-121.

[7] Keys A. (1986). The diet and 15-years death rate in the seven countries study. *Am. J. Epidemiol.* 124 (6). 903-915.

[8] Colomer R. & Menéndez J. A. (2006). Mediterranean diet, olive oil and cancer. *Clin. Tranl. Oncol.* 8. 15-21.

[9] Martínez-González M. A., Salas-Salvado J., Estruch R. & Corella D. (2015). Benefits of the mediterranean diet: Insights from the PREDIMED Study. *Prog. Cardiovasc. Dis.* 58. 50-60.

[10] Trichopoulou A., Costacov T., Bamia C. & Trichopoulus D. (2003). Adherence to a mediterranean diet and survival a Greek population. *N. Engl. J. Med.* 348. 2599-2608.

[11] Huang C. L. & Sumpio B. E. (2008). Olive oil the mediterranean diet and cardiovascular health. *J. Am. Coll. Surg.* 207. 407-416.

[12] Marlow G., Ellett S., Ferguson I. R., Zhu S., Karunasinghe N., Jesuthasan A. C., Han D., Fraser A. C. & Ferguson L. R. (2013). Transcriptomics to study the effect of a mediterranean inspired diet on inflamamation in Crohn′s disease patientss. *Hum. Genom.* 7. 24-30.

[13] Visioli F. & Bernardini E. (2011). Extra virgin olive oil′s polyphenols: Biological activities. *Curr. Pharm. Des.* 17. 786-804.

[14] Arslan D., Karabekir Y. & Schreiner M. (2013). Variations of phenolic compounds, fatty acids and some qualitative characteristics of Sariulak olive oil as induced by growing area. *Food Research Internation.* 54. 1897-1906.

[15] Ricciutelli M., Marconi S., Boarelli M. Ch., Caprioli G., Sagratini G., Ballini R. & Fiorini D. (2017). Olive oil polyphenols: A quantitative method by high-performance liquid-chromatography-diode-array detection for their determination and the assessment of the related health claim. *Journal of Chromatography A.* 1481. 53-63.

[16] Peres F., Martins L. L., Mourato M., Vitorino C. & Ferreira-Dias S. (2016). Bioactive compounds of Portuguese virgin olive oils. Discriminate cultivar and ripening stage. *J. Am. Oil Chem. Soc.* 99. 1137-1147.

[17] FDA Food and Drug Administration. USA. (2004). Docket no 2003Q-0559. *Monounsaturated fatty acids from olive oil and coronary heart disease.*

[18] Frankel E. N. (2011). Nutritional and biological properties of extra virgin olive oil. *J. Agric. Food Chem.* 59. 785-792.

[19] Visioli F. (2015). Xanobiotics and human health: A new view of their pharma-nutritional role. *PharmaNutrition*. 3. 60-64.

[20] La Vecchia C. (2004). Mediterranean diet and cancer. *Public Health Ntr*. 7. 965-968.

[21] Bernardini E. & Visioli F. (2017). High quality, good health: The case for olive oil. *Eur. J. Lipid Sci. Technol.* 119. 1500525. 1-9.

[22] Caramia G., Gori A., Valli E & Cerretani L. (2012). Virgin olive oil in preventive medicine: From legend to epigenetics. *Eur. J. Lipid Sci. Technol.* 114. 375-388.

[23] Romero C., Ruíz-Méndez V. & Brenes M. (2016). Bioactive compounds in virgin olive oil of the PDO Montoro-Adamuz. *J. Am. Oil Chem. Soc.* 93. 665-672.

[24] Camargo A., Delgado-Lista J., García-Ríos A., Cruz-Teno C., Yubero-Serrano E. M., Pérez-Martínez P., Gutierrez-Mariscal F. M., Lora-Aguilar P., Rodríguez-Cantalejo F. & Fuentes-Jiménez F. (2012). Expresion proinflamatoria proatherogenic genes is reduced by the Mediterranean diet in elderly people. *British J. Nutr*. 108. 500-508.

[25] Bendini A., Cerretani L., Carrasco-Pancorbo A., Gómez-Caravaca A. M., Segura-Carretero A., Fernández-Gutiérrez A. & Lercker G. (2007). Phenolic molecularss in virgin olive oils: a survey of their sensory properties, health effects, antioxidant activity and analytical methods. An orverview of the last decade. *Molecules*. 12. 1679-1683.

[26] Oliveras-López M. J., Berná G., Jurado-Ruíz E., López-García de la Serrana H. & Martin F. (2014). Consumption of extra virgin olive oil rich in phenolic compounds has beneficial antioxidant effects in healthy human adults. *J. Funct. Foods*. 10. 475-484.

[27] Covas M. I., de la Torre R. & Fito M. (2015). Virgin olive oil: A key food for cardiovascular risk protection. *Br. J. Nutr*. 113. 519-528.

[28] Jones A (2011). Can mediterraneann type diet prevent Parkinson's disease. *Neurol. Rev*. 19. 1. 21.

[29] López S., Bermudez B., Monserrat-de la Paz S., Jaramillo S., Aia R. & Muriana F. J. G. (2016). Virgin olive oil and hypertension. *Current Vascular Pharmacology*. 14. 323-329.

[30] Aparicio R. & Luna G. (2002). Characterization of monovarietal virgin olive oils. *Eur. J. Lipid Sci. Technolo*. 104. 614-627.

[31] López-Cortés I. & Salazar D. M. (2006). *Variedades de olivo y composición de sus aceites en el Oeste del Mediterráneo. [Composition and varieties of olive crop on the Mediterrannean west]* Phytoma-España Ed. Valencia. Spain, 443 pp.

[32] Quintero-Flórez A., Sinausa L., Sánchez-Ortiz A., Beltrán G. & Perona J. S. (2015). The fatty acid composition of virgin olive oil from different cultivars is determinant for foam cell formation by macrophages. *J. Agric. Food Chem*. 63. 6731-6738.

[33] Bouarroudj K., Tamendjari A. & Larbat R. (2016). Quality, composition antioxidant activity of Algerian wild olive (*Olea europaea* L. sbsp. *Oleaster*) oil. *Industrial Corps and Products*. 83. 484-491.

[34] Ceci L. N. & Carelli A. A. (2010). Relation between oxidative stability and composition in Argentinian olive oils. *J. Am. Oil Chem*. Sec. 87. 1189-1197.

[35] Fares R., Bazzi S., Baydoun S. E. & Abdel-Massih R. M. (2011). The antioxidant and anti-proliferative actibity of the Lebanese *Olea europaea* extract. *Plant Foods Hum. Nutr*. 66. 58-63.

[36] Moreno J. J. (2013). Effect of olive oil minor componentes on oxidative stress and arachidonic acid mobilization and metabolism by macrophages. *Free Radic. Biol. Med.* 35. 1073-1081.

[37] Barrajon-Catalán E., Taamalli A., Quirantes-Piné R., Roldan-Segura C., Arraez-Roman D., Segura-Carretero A., Micol V. & Zarrouk M. (2015). Differential metabolomic analysis of the potential antiproliferative mechanism of olive leaf extract on the JIMT-1 breast cancer cell line. *J. Pharm. Biomed. Anal*. 105. 156-162.

[38] López-Cortés I., Salazar-García D. C., Velázquez-Martí B. & Salazar D. M. (2013). Chemical characterization of traditional

varietal olive oils in East of Spain. *Food. Res. Int.* 54 (2). 1934-1940.

[39] Kendall M., Batterham M., Callahan D. L., Jardine D., Prenzler P. D., Robards K. & Ryan D. (2012). Randomized controlled study of the urinary excretion of biophenols following acute and cronic intatie of olive leaf supplements. *Food Chem.* 130. 651-659.

[40] Yuan J. J., Wang C. Z., Ye J. Z., Tao R. & Zhang Y. S. (2015). Enzymatic hydrolysis of oleuropein from *Olea europea* (olive) leaf extract and antioxidant activities. *Molecules*. 2903-2921.

[41] Jemai H., El Feki A. & Sayadi S. (2009). Antidiabetic and antioxidant effects of hydroxytyrosol and oleuropein from olive leaves in alloxan-diabetic rats. *J. Agric. Food Chem.* 57. 8798-8804.

[42] Fitó M., De la Torre R. & Covas M. I. (2007). Olive oil and oxidative stress. *Mol. Nutr. Food Res.* 51. 1215-1224.

[43] Weinbrenner T., Fitó M., de la Torre R., Saez G. T., Rijken P., Tormos C., Coolen S., Albaladejo M. F., Abanades S., Schroder H., Marrugat J. & Covas M. I. (2004). Olive oils high in phenolic compounds modulate oxidative/antioxidative status in men. *J. Nutr.* 134. 2314-2321.

[44] Boss A., Bishop K. S., Marlow G., Barnett M. P. G. & Ferguson L. R. (2016). Evidence to support the anti-cancer effect of olive leaf extract and future directions. *Nutrients*. 8. 513-547.

[45] Mijatovic S. A., Timotijevic G. S., Miljkovic D. M., Radovic J. M., Maksimovic-Ivanic D. D., Dekanski D. P. & Stosic-Grujicic S. D. (2011). Multiple antimelanoma potential of dry olive leaf extract. *Int. J. Cancer*. 128. 1955-1965.

[46] Pereira A. P., Ferreira I. C., Marcelino F., Valentio P., Andrade P. B., Seabra R., Estevinho L., Bento A. & Pereira J. A. (2007). Phenolic compounds and antimicrobial activity of olive (*Olea europaea* L. cv, Cobrançosa) leves. *Molecules*. 12. 1153-1162.

[47] Lee O. H. & Lee B. Y. (2010). Antioxidant and antimicrobial activities of individual and combined phenolics in *Olea europaea* leaf extract. *Bioresour. Technol.* 101. 3751-3754.

[48] El S. N. & Karakaya S. (2009). Olive tree (*Olea europaea* L.) leaves: Potential beneficial effects on human health. *Nutr. Rev*. 67. 632-638.

[49] Benavente-García O., Castillo J., Lorente J. & Ortun A. (2000). Antioxidant activity of phenolics extracted from *Olea europaea* L. leaves. *Food Chem*. 68. 457-462.

[50] Guinda A., Pérez-Camino M. C. & Lanzón A. (2004). Supplementation of oils with oleanolic acid from the olive leaf (*Olea europaea* L.). *Eur. J. Lipid Sci. Technol*. 106. 22-26.

[51] De Marino S., Festa C., Zollo F., Nini A., Antenucci L., Raimo G. & Iorizzi M. (2014). Antioxidant activity and chemical components as potential anticancer agents in the olive leaf (*Olea europaea* L. cv. Leccion) decoction. *Anticancer Agents Med. Ch*em. 14. 1376-1385.

[52] Milanizadeh S., Bigdeli M. R., Rasoulian B. & Amani D. (2014). The effects of olive leaf extract on antioxidant enzymes activity and tumor growth in breast cancer. *Thrita*. 3. 127-135.

[53] Ramírez E., Medina E., Brenes M. & Romero G. (2014). Endogenous enzymes involved in the transformation of oleuropein in Spanish table olive varieties. *J. Agric. Food Chem*. 62. 9569-9575.

[54] Pereira J. A., Pereira A. P. G., Ferreira I. G. F. R. & Valentap P. (2006). Table olives from Portugal, phenolic compounds, antioxidant potential and antimicrobial activity. *J. Agric. Food Chem*. 54. 8425-8431.

[55] Wilt T. T., MacDonald R. & Ishani A. (1999). Betasitosterol for the treatment of benign prostatic hyperplasia a systematic review. *Br. J. Urol*. 83. 976-983.

[56] Piroddi M., Albini A., Fabiani R., Giovannelli L., Luceri C., Natella F., Rosignoli P., Rossi T., Taticchi A., Servili M & Galli F. (2017). Nutrigenomics of extra-virgin olive oil. Areview. *I. Union Biochem. and Mol. Biology*. 43. (1). 17-41.

[57] Carrillo C., Cavia M. D. & Alonso-Torre S. R. (2012). Antitumor effect of oleic acid, mechanisms of action, a review. *Nutr. Hosp*. 27. 1860-1865.

[58] Varela L. M., Ortega-Gómez A., López S., Abia R., Muriana F. J. & Bermudez B. (2013). The effects of dietary fatty acid on the postprandial triglyceride-rich lipoprotein/apo B48 receptor axis in human monocyte/macrophague cells. *J. Nutr. Biochem.* 24. 2031-2039.

[59] Yan J., Shen Y., Wang Y., Luan X., Gou M. & Cong L. (2016). The effects of different extraction methods on the physicochemical properties and antioxidant activity of *Amygdalus* seed oil. *Jour. Appl. Botany and Food Quality*. 89. 135-141.

[60] Parry J., Lan S., Luther M., Zhon K., Peter M. & Whittaker P. (2005). Faty acid composition and antioxidant properties of cold-pressed marionberry, boysenberry, red raspberry and blueberry seed oil. *J. Agr. Food Chem.* 53. 566-573.

[61] Cabello-Moruno R., Perona J. S. & Ruíz-Gutierrez V. (2007). Influence of minor componentes of olive oils on the composition and size of TRLs and on macrophage receptors involved in foam cell formation. *Biochem. Soc. Trans.* 35. 470-471.

[62] Moore E. H., Napolitano M., Avell M., Bejta F., Sucklingl E., Bravo E. & Botham K. M. (2004). Protection of chylomicron remnants from oxidation by incorporation of probucol into the particles enhances their uptake by human macrophages and increases lipid accumulation in the cells. *Eur. J. Biochem.* 271. 2471-2427.

[63] Al-Ramawi F., Odeh I., bisher A., Abbadi J. & Qabbajeh M. (2014). Effect of geographical region and harvesting date on antioxidant activity, phenollic and flavanooid content of olive leaves. *J. Food Nutr. Res.* 2. 925-930.

[64] Bakhouche A., Lozano-Sánchez J., Beltrán-Debón R., Joven J., Segura-Carretero A. & Fernández-Gutiérrez A. (2013). Phenolic characterization and geographical classification of commercial Arbequina extra virgin olive oils produced in southern Catalonia. *Food Res. Int.* 50. 401-408.

[65] Gimeno E., Castellote A. L., Lamuela-Raventos R. M., de la Torre M. C. & López-Sabater M. C. (2002). The effect of harvest and

extraction methods on the antioxidation content (phenolic, α-tocoferol and β-carotene) in virgin olive oil. *Food Chem.* 78. 207-211.

[66] Hannachi H., Nasri N., el Falleh W., Tlili N., Faerchichi A. & Msallem M. (2013). Fatty acid, sterols, polyphenols and chlorophylls of olive oils obtained from Tunisian wild olive trees (*Olea europaea* L. var. sylvestris). *Int. J. Food Prop.* 16. 1271-1278.

[67] Hioki H., Miura T., Miyashita Y., Ebisawa S., Motoki H., Izawa A., Koyama J. & Ikeda U. (2016). Circulating eicosapentanoic acid to oleic acid ratio and risk for cardiovascular events in patients with coronary artery disease. *IJC. Metabolic & Endocrine*. 10. 1-6.

[68] Schwingshackl L. & Hoffmann G. (2014). Monounsatured fatty acids, olive oil and health status: A systematic review and meta-analysis of cohort studies. *Lipids Health Dis.* 13. 154-159.

[69] Mozaffarian D. (2016). Dietary and policy priorities for cardiovascualr disease. Diabetes and obesity: A comprehensive review. *Circulation*. 133. 187-225.

[70] Weinstock-Guttman B., Baier M., Park Y., Feicter J., Lee-Kwen P. & Gallagher E. (2005). Low fat dietary intervention with omega3 fatty acid supplementation in multiple sclerosis patients. *Prostaglandina Leukot Essent Fatty Acids*. 73. 397-404.

[71] Solfrizzi V., Panza F., Torres F., Mastroianni F. Del Parigi A. & Venezia A. (1999). High monosaturated fatty acids intake protects against age-related cognitive decline. *Neurology*. 52. 1563-1569.

[72] Facchinetti F., Dawson V. L. & Dawson T. M. (1998). Free radicals as mediators of neuronal injury. *Cell. Mol. Eurobiol.* 18. 667-682.

[73] García-González D., Rowoeo N. & Rawn A. (2010). Comparative study of virgin olive quality from varieties cultivated in Chilie and Spain. *J. Agric. Food Chem.* 58. 12899-12905.

[74] Petrakis P., Agiomyrgianaki A., Christophoridou S., Spyros A. & Dais P. (2008). Geographical characterization of Greek virgin olive oils (cv. Koroneiki) using 1H and 31P NMR fingerprinting with

canonical discrimiannt analysis and classification binary trees. *J. Agric. Food Chem.* 56. 3200-3207.

[75] Romero M. P., Tovar M. J., Ramo T. & Motilva M. J. (2003). Effect of crop season onthe composition of virgin olive oil with protected designation of origin Les Garrigues. *J. Am. Oil Chem. Soc.* 80. 423-431.

[76] Uluata S., Altuntas Ü. & Ozçelik B. (2016). Biochemical characterizationof Arbequina extra virgin olive oil produced in Turkey. *J. Am. Oil Chem. Soc.* 93. 617-626.

[77] Beltrán G., Aguilera M. P., Del Río C., Sánchez S & Martinez L. (2005). Influence of fruit ripening process on the natural antioxidant content of Hojiblanca virgin ollive oils. *Food Chem.* 89. 207-215.

[78] Fernández-Cuesta A., León L., Velasco L. & De la Rosa R. (2013). Changes in squalene and sterols associated with olive maturation. *Food Res. Int.* 54. 1885-1889.

[79] Jiménez B., Sánchez-Ortiz A., Lorenzo M. L. & Rivas A. (2013). Influence of fruit ripening on agronomic parameters, quality indices, sensory attributes and phenolic compunds of Picudo olive oils. *Food Res. Int.* 54. 1860-1867.

[80] Lazzez A., Perri E., Anna C. M., Cossentini M., Khlif M. & Cossentini M. (2008). Influence of olive maturity stage and geographical origin on some minor componentes in virgin olive oil of the Chemlali variety. *J. Agric. Food Chem.* 56. 982-988.

[81] Katsoyannos E., Batrinou A., Chatzilazarou A., Bratakos S. M., Stamatopoulos K. & Sinanoglou V. I. (2015). Quality parameters of olive oil from stoned and non stoned Koroneiki and Megaritiki Greek olive varieties at different maturity levels. *Grasas y Aceites.* 66 (1). 8 pp.

[82] Carrasco-Pancorbo A., Cerretani I., Bendini A., Segura-Carretero A., Gallina-toschi T. & Fernández-Gutierrez A. (2005). Analytical determiantion of polyphenols in olive oils. *J. Sep. Sci.* 28. 837-258.

[83] Petroni A., Balsecvich M., Salami M., Papini N., Montedoro G. G. & Galli C. (1995). Inhibition of platelet aggregation and ercosanoid

production by phenolic componentes of olive oil. *Thromb. Res.* 78. 151-160.

[84] D´Archivio M., Filesi C., Vari R., Scazzocchio B. & Masella R. (2010). Bioavailability of the polyphenols: Status and controversies. *Int. J. Mol. Sci.* 11. 1321-1324.

[85] Ciriminna R., Meneguazo F., Fidalgo A., Ilharco L. M. & Pagliaro M. (2016). Extraction, benefits and valorization of olive polyphenols. *Eur. J. Lipid Sci. Technol.* 118. 5003-5011.

[86] Flores A., Isabel M., Romero-González R., Frenich G., Vidal A. &Martínez L. J. (2012). Analysis of phenolic compounds in olive oil by solid phase extration and ultra high performance liquid chromatrography tandem mass spectrometry. *Food Chem.* 134. 2465-2472.

[87] Cicerale S., Lucas L. & Keast R. (2010). Biological activities of phenolic compounds present in virgin olive oil. *Int. J. Mol. Sci.* 11. 458-479.

[88] Bogani P., Galli C., Villa M. & Visioli F. (2007). Postprandial anti-inflammatory and antioxidant effects of extra virgin olive oil. *Atherosclerosis.* 190. 181-186.

[89] Franco M. N., Galeano-Díaz T., López O., Fernández-Balaños J. G., De Miguel C., Gil M. V. & Martín-Vertedor D. (2014). Phenolic compounds and antioxidant capacity of virgin olive oil. *Food Chem.* 163. 289-298.

[90] Visoli F., Poli A. & Gall C. (2002). Antioxidant and other biological activities of phenols from olives and olive oil. *Med. Res. Rev.* 22. 65-75.

[91] Vissers M. N., Zock P. L. & Katan M. B. (2004). Bioavalilability and antioxidant effects of olive oil phenols in humans: a review. *Eur J. Clin. Nutr.* 58. 955-965.

[92] Cardeno A., Sánchez-Hidalgo M. & Alarcón-De la Lastra C. (2013). An up-date of olive oil phenols in inflammation and cancer: Molecular mechanisms and clinical implications. *Curr. Med. Chem.* 20. 4758-4776.

[93] Owen R. W., Mier W., Giacossas A., Hull W. E., Spiegelhalder B. & Bartsch H. (2000). The antioxidant/anticancer potential of phenolic compounds isolated from olive oil. *Eur. J. Cancer.* 36. 1235-1247.

[94] Bieniek J., Childress C., Swatski M. D. & Yang W. (2014). COX-2 inhibitors arrest prostate cancer cell cycle propression by down-regulation of kinelochore centromere proteins. *Prostate*. 74. 999-1011.

[95] Peng L., Zhou Y., Wang Y., Mou H. & Zhao Q. (2013). Prognostic significance of COX-2 inmunohistochemical expression in coleorectal cancer: A meta analysis of the literature. *PlosOne*. 8 (3).

[96] Scoditti E., Calabriso N., Masiaro M., Pellegrino M., Storelli C., Martínez G., De Caterina R. & Carluccio M. A. (2012). Mediterranean diet polyphenols reduce inflammatory angiogenesis through MMP-9 and COX-2 inhibition in human vascular endothelial cells: A potentially protective mechanism in atherosclerotic vascular diseases and cancer. *Arch. Biochem. Biophys*. 527. 81-89.

[97] De la Puerta R., Ruíz Gutierrez V. & Hoult J. R. (1999). Inhibition of leukocyte 5-lipoxygenase byphenolics from virgin olive oil. *Biochem. Pharmacol.* 57. 445-449.

[98] Khalatbary A. R. (2013). Olive oil phenols neuroprotection. *Nutritional Neuroscience*. 16 (6). 243-249.

[99] Rosillo M. A., Alcaraz M. J., Sánchez-Hidalgo M., Fernández-Bolaños J. G., Alarcón de la Lastra C. & Ferrándeiz M. L. (2014). Anti-inflammatory and joint protective effects of extra-virgin olive oil polyphenol extract in experimental arthritis. *J. Nutritional Biochemistry*. 25. 1275-1281.

[100] Corona G., Tzounis X., Dessl A. M., Deiana M., Debmam E. S., Visioli F. & Spencer J. P. E. (2006). The fate of olive oil polyphenols in the gastrointestinal tract: Implications of gastric and colonic microflora dependent biotransfromation. *Free Radic Res.* 40. 647-658.

[101] De la Puerta R., Domínguez M. E. M., Ruíz-Gutiérrez V., Flavill J. A. & Hoult J. R. S. (2011). Effects of virgin olive oil phenolics on scavenging of reactive nitrogen species and upon nitrogenic neurotransmission. *Life Sci.* 69. 1213-1222.

[102] Tasset I., Pontes A. J., Hinojosa A. J., de la Torre R. & Túnez J. (2011). Olive oil reduces oxidative damage in a 3-nitropropionic acid induced huntington's disease like rat model. *Nutritional Neuroscience*. 14. 106. 11.

[103] Bayran B., Esatbeyoglu T., Schulze N., Ozcelik B., Frank J. & Rimbach G. (2012). Comprehensive analysis of polyphenols in 55 extra virgin olive oils by HPLC-FCD and their correlation with antioxidant activities. *Plant Foods Hum. Nutr.* 67. 326-336.

[104] García A., Brenes M., García P., Romero C., & Garrido A. (2003). Phenolic content in commercial olive oils. *Eur. Food Res. Technol.* 216. 520-525.

[105] Visolo F., Bellono G. & Galli C. (1998). Free radical scavenging properties of olive oil polyphenols. Biochem. *Biophys. Res. Com.* 247. (1). 60-64.

[106] Servili M., Selvaggini R., Esposto S., Taticchi A., Montedoro G. & Morozzi G. (2004). Health and sensory properties of virgin olive oil hydrophilic phenols: agronomic and technological aspects of production that affect their occurrence in the oil. *J. Chromatogr.* 1054. 113-127.

[107] González-Santiago M., Fonolla J. & López-Huertas E. (2010). Human absorption of a supplement containing purified hydroxytyrosol, a nautral antioxidant from olive oil and evidence for its transient association with low-desnity lipoproteins. *Pharmacol. Res.* 61. 364-370.

[108] Aunnon-Calles D., Giordano E., Bohnenberger S. & Visioli F. (2013). Hydroxytyrosol is not genotoxic invitro. *Pharmacol. Res.* 74. 87-93.

[109] Fernández-Bolanos J. G., López O., Juan F., Fernández-Bolanos J. & Rodríguez-Gutierrez F. (2008). Hydroxytyrosol and derivates Isolation synthesis and biological propierties. *Curr. Org. Chem.* 135. 1475-1479.

[110] Granados-Principal S., Quiles J. L., Ramírez-Tortosa C. L., Sánchez-Rovira P. & Ramírez-Tortosa M. C. (2010). Hydroxytyrosol from laboratory investigations to future clinical riels. *Nutr. Rev*. 68. 191-206.

[111] Zhang X., Cao J. & Zhong L. (2009). Hydroxytyrosol inhibits pro-inflamatory cytokines iNOs and COX-2 expression in human monocytes cells. *Naunyn Schmied. Arch. Pharmacol. 379*. 581-586.

[112] Scoditti E., Nestola A., Massaro M., Calabriso N., Storelli C., De Caterina R. & Carluccio M. A. (2014). Hydroxytyrosol suppresses MMP-9 and COX-2 activity and expression in activated human monocytes via PKCα and PKCβ_1 inhibition. *Atherosclesosis. 232*. 17-24.

[113] Fuccelli R., Fabiani R., Sepporta M. V. & Rosignoli P. (2015). The hydroxytyrosol dependent increase of TNF-α in LPS-activated human monocytes is mediated by PGE-2 and adenylate ciclasa activation. *Toxicol. Vitro. 29*. 933-937.

[114] Silva S., Sepodes B., Rocha J., Direito R., Fernándes A., Brites D., Freitas M., Fernándes E., Bronze M. R. & Figuiera M. E. (2015). Protective effects of hydroxytyrosol supplemented refined olive oil in animal models of acute inflammation and rheumatoid arthritis. *Journal of Nutritional Biochemistry*. 26. 360-368.

[115] Zrelli H., Matsuoka M., Kitazaki S., Zarrouk M. & Miyazaki H. (2011). Hydroxytyrosol reduces intracellular reactive oxygen specieslevels invascular endothelial cells by upregulating catalase expression through the AMPK-FOXO 3a pathway. *Eur. J. Pharmacol*. 660. 275-282.

[116] Giordano E., Davalos A., Nicod N. & Visioli F. (2014). Hydroxytyrosol attenuates -tunicamycin induced endoplasmic reticulum stress in human hepatocarcinoma cells. *Mol. Nutr. Food Res*. 58. 954-962.

[117] Bernini R., Gilardini M. S., Merendino N., Romani A. & Velotti F. (2015). Hydroxytyrosol derived compounds: a basis for the creation of new pharmacological agents for cancer prevention and therapy. *J. Med. Chem.* 58. 9089-9107.

[118] Omar S. H. (2010). Oleuropein in olive and its pharmacological effects. *SciPharm.* 78. 133-154.

[119] Nekooeian A. A., Khalili A. & Khosravi M. B. (2015). Oleuropeina offers cardioprotection in rats with simultaneous type diabetes and renal hypertension. *Indian J. Pharmacol.* 46. 398-403.

[120] Elamin M. H., Daghestani M. H., Omer S. A., Elobeid M. A., Virk P., Al-Olayan E. M. Hassan Z. K., Mohammed O. B. & Aboussekhra A. (2013). Olive oil oleuropein has anti-breast cancer properties with higher efficiency on ER-negtive cells. *Food Chem. Toxicol.* 53. 310-316.

[121] Han J., Talorete T. P. N., Yamada P. & Isoda H. (2009). Anti-proliferative and apoptotic effects of oleuropein and hydroxytyrosol on human breast cancer MCF-7 cell. *Cytotechnology.* 59. 45-53.

[122] Sepporta M. V., Fuccelli R., Rosignoli P., Ricci G., Servili M., Murozzi G. & Fabiani R. (2014). Oleuropein inhibits tumor growth and metastases dissemination in ovariectomised nude mice with MCF-7 human breast tumour xenografs. *J. Funct. Foods.* 8. 269-273.

[123] Yao J., Wu J., Yang X., Yang J., Zhang Y. & Du L. (2014). Oleuropein induced apoptosis in Hela cells via mitochondrial apopototic cascade associated with activation of the C. JunNH2-terminal kinase. *J. Pharmacol. Sci.* 125. 300-311.

[124] Jukanti A. K., Gaur P. M., Gowda C. L. L. & Chibbar R. N. (2012). Nutritional quality and health benefits of chickpea (*Cicer arietinum* L.). *British J. Nutrition.* 108. 512-526.

[125] Kritchevsky D. & Chen S. D. (2005). Phytosterols health benefits and potential ocncerns: A review. *Nutr. Res.* 25. 413-428.

[126] Kyçyk O., Aguilera M. P., Gaforio J. J., Jiménez A. & Beltrán G. (2016). Sterol compositionof virgin olive oil of forty-three olive

cultivars from the world collection olive germplasm Bank of Cordoba. *J. Sci. Food Agric*. 96. 4143-4150.

[127] Bouic P. J. D. (2002). Sterols and sterolins: new drugs for the immune system. *Drug Discov. Today*. 7. 775-778.

[128] Awad A. B., Downie A., Fink C. S. & Kim U. (2000). Dietary phytosterol inhibits the growth and metastasis of MDA-MB-231 human breast cancer cells grown in SCID mice. *Anticancer Res*. 20. 821-824.

[129] Klingber S., Ellegard L., Johansson J., Jansson J. H., Hallmans G. & Winkvist A. (2013). Dietary intake of naturally occurring plant sterols is related to a lower risk of a first myocardial infraction i men but not in women in Northern Swedem, *J. Nutr. 143*. 1630-1635.

[130] Law M. R. (2000). Plant sterol and stanol and health. *West J. Med*. 173. 43-47.

[131] Klipper K. T., Hiltl D. M. & Schipp B. A. (1997). A multicentric placebo controlled double blind clinial trial of β-sitosterol (phytosterol) for the treatment of benign prostatic hyperplasia. *Br. J. Urol*. 80. 427-432.

[132] Awad A. B., Downie A. C. & Fink C. S. (2000). Inhibition of growth and stimulation of apoptosis by beta-sitosterol treatment of MDA-MB-231 human breast cancer cells inculture. *Int. J. Mol. Med*. 5. 541-545.

[133] Portarena S., Baldacchini C. & Brugnoli E. (2017). Geographical discrimination of extra virgin olive oils from the Italian coasts by combining stable isotope data and carotenoid content within a multivariate analysis. *Food Chem*. 215. 1-6.

[134] Sáez G. & Puig-Parellada P. (1999). *Radicales libres y estrés oxidativo en biomedicina. Importancia y utilidad de los antioxidantes en la prevención de procesos fisiopatológicos relacionados. [Free radicals and oxidative stress in biomedicine. Importance and usefulness of antioxidants in the prevention of pathophysiological processes related]* Universitat de Valencia. FVEA Ed. Valencia. Spain. 177 pp.

[135] Virtamo J., Taylor P. R., Kontto J., Männistö S., Utriainen M., Weinstein S. J., Huttunem J. & Albanes D. (2014). Effects of α-tocopherol and β-caroteno supplementation on cancer incidence and mortality: 18 year posintervention follow-up of the α-tocopherol, β-carotene cancer preention study. *Int. J. Cancer*. 135. 178-185.

[136] Beltrán G., Jiménez A., Del Río C., Sánchez S., Martínez L., Uceda M. & Aguilera M. P. (2010). Variability of vitamin E in virgin olive oil by agronomical and genetic factors. *J. Food Comp. Anal. 23*. 633-639.

[137] González M. (2011). *Aceite de oliva. Beneficios para la salud. [Olive oil. Health Benefits]*. Faeditorial Ed. Alcalalí La Real. Jaén. Spain. 136 pp.

[138] Caramia G. & Fanos V. (2007). Mediatori lipidici, infezioni e inflammazioni: Evoluzione delle conocenze e prospettive terapeutiche. *Giorn. It. Inf. Ped*. 9. 15-27.

[139] Cotran R. S. & Munro J. M. (1988). *Patogenia de la aterosclerosis*. In Grundy S. M. and Bearn. *Colesterol y aterosclerosis. Nuevas posibilidades terapéuticas*. [*Cholesterol and Atherosclerosis. New therapeutic possibilities*]. Merck Ed. Nueva Jersey. USA.

[140] Hugen C. L. & Sumpio B. E. (2008). Olive oil the mediterranean diet and cardiovascular health. *J. Am. Coll. Surg. 207*. 407-416.

In: Olive Oil
Editors: T. Fritjof and B. Henning ISBN: 978-1-53612-563-4

Chapter 2

OLIVE OIL PHYTOSTEROLS AND HUMAN HEALTH

Hedia Manai-Djebali [*] ***and Imen Oueslati***
Laboratory of Olive Tree Biotechnology,
Centre of Biotechnology of Borj-Cédria, Hammam-Lif, Tunisia

ABSTRACT

The chemical composition of olive oil has been traditionally clustered into major and minor compounds; the formers are, in a large part, responsible for the olive oil main characteristics while the latters are markers for their peculiarities. The minor fraction also known as unsaponifiable fraction is rich in minor metabolites such as sterols, triterpene dialcohols, phenolic compounds, tocopherols, hydrocarbons, pigments, etc. Phytosterols make up the greatest proportion of the unsaponificable fraction and are the most used for authenticity and discrimination purposes. During the last years, sterols have received particular attention because of their capability to lower serum cholesterol levels in the human body resulting in significant reduction in the risk of cardiovascular disease. Furthermore, they are considered as anti-inflammatory, anti-bacterial, antioxidant and anti-cancer agents. The

[*] Corresponding Author Email: hedia_manai@yahoo.fr; Hedia.Manai@.cbbc.rnrt.tn.

analysis of the olive oil sterol contents is usually carried out according to the official method provided by the EU and the IOC. Several authors highlighted that the amount and the composition of sterols depend on many factors such as olive variety, ripening degree, oil extraction procedures, agronomic and climatic conditions.

Keywords: olive oil, phytosterols, authenticity, discrimination, health benefits

Introduction

Plant sterols, also called phytosterols, are an essential component of the membranes of all eukaryotic organisms. Their function appears to be controlling membrane fluidity and permeability. In addition, sterols are an essential part of cellular signaling systems, as well as being general precursors in hormone biosynthesis (Piironen et al., 2000).

Sterols can occur in vegetable oils either in free form or esterified with fatty acids. Since both fractions have different compositions, the combined determination of these two classes of compounds provides a more informative approach to check vegetable oils authenticity (Mariani et al., 1999).

Vegetable oils and oil-based products are the richest dietary sources of phytosterols, followed by cereal grains, cereal-based products, and nuts (Piironen et al., 2000).

Virgin olive oil (VOO) is one of the oldest known vegetable oils and is extracted from the fruit of *Olea europaea* L. It is regarded by consumers around the world as one of the healthiest vegetable oils, because of its role in disease prevention.

Olive oil naturally contains several minor compounds such as hydrocarbons, sterols, aliphatic alcohols, phenolic compounds, tocopherols, pigments and flavor components.

Sterols are the main constituents of the unsaponifiable fraction and their content corresponds to around 20% of the total unsaponifiable matter of olive oil (Matos et al., 2007; Lukić et al., 2013).

Compositional analysis of the olive oil sterol fraction can be used to assess the degree of purity of the oil and the absence of other vegetable oils. Research has shown that each olive oil has a characteristic sterol profile which can be used for checking its genuineness (Salvador et al., 1998; Gutiérrez et al., 2000).

The sterol determination also permits the characterization of the olive oil category: extra virgin, virgin, refined, etc. (IOC, 2016; EU Regulation, 2016). Example for extra virgin olive oil, the EU regulation and IOC established a minimum sterol content of 1000 mg kg^{-1} and a minimum concentration of 93% of apparent β-sitosterol, which includes both β-sitosterol and Δ-5-avenasterol together with other minor sterols formed by the degradation of β-sitosterol (sitostanol, Δ-5,24-stigmastadienol and clerosterol) (IOC, 2016; EU Regulation, 2016)

Virgin olive oil has moderate sterol content, typically between 1000 and 2000 mg kg^{-1}, compared to larger contents in seed oils such as soybean (2290 to 4590 mg kg^{-1}), sunflower (3740 to 7250 mg kg^{-1}), rapeseed (4130 to 9790 mg kg^{-1}) and corn (8090 to 15,570 mg kg^{-1}) (Fernandez-Cuesta et al., 2013).

The determination of the sterol composition is very important in the characterization of olive oils as well as in the quality control, since they can be used to determine possible frauds. Additionally, it has been reported that phytosterols have anti-inflammatory, antibacterial, antifungal, antiulcerative, antitumoral and antioxidant activities (Williamson, 1988; Awad and Fink, 2000; Saeidnia et al., 2014).

The health aspects of β-sitosterol, major sterol of olive oil, have also been reported in several studies. They refer mainly to the reduction of cholesterol levels and the prevention of many diseases and various types of cancer (colon, prostate, and breast) (Saeidnia et al., 2014).

The European Foods Safety Authority (EFSA) recommends consuming about 1.5–2.4g of phytosterols/day and/or stanols, in order to reduce blood cholesterol (EFSA, 2008). Furthermore, the Food and Drug Administration (FDA) has approved the role of foods containing phytosterols inside a low saturated fat in reducing the risk of heart disease, especially consumption of at least 1.3g of phytosterols/day, twice a day (FDA, 2010).

Phytosterols are also components for the oil stability since at high temperature they act as antioxidants; they are inhibitors of polymerization reactions (Velasco and Dobarganes, 2002).

It has been demonstrates by several studies that olive oil total sterol content and profile can be affected by many factors such as variety, agronomic and climatic conditions, fruit quality and maturity, oil extraction and refining procedures and storage conditions prior to oil extraction (Cañabate-Díaz et al., 2007; Fernandez-Cuesta et al., 2013; Yorulmaz et al., 2014; Konuskan and Mungan, 2016).

The aim of this chapter is to present an overview the importance of the phytosterols in olive oil and list their human health benefits. For their importance, these components are used to accomplish authenticity of olive oil, traceability of geographical origin and cultivar identification.

1. Olive Oil Phytosterols

It must be pointed out that there are official methods developed for determining total and individual contents of phytosterols in olive oil. The official methods were provided by the International Olive Council (IOC) and the European Union (EU).

Currently, the most common methods for the determination of olive oil phytosterols involve alkaline hydrolysis (saponification), followed by extraction of the non-saponifiables, cleanup of the extract, separation and derivatisation of the sterols, quantification of the sterol derivatives by Gas Chromatography (GC) using a capillary column.

High performance liquid chromatography (HPLC) with different detection systems has also been used to identify and quantify phytosterols in olive oils (Cunha et al., 2006; Cañabate-Díaz et al., 2007; Zarrouk et al., 2009; Lerma-García et al., 2009; Xianga et al., 2017).

Sterols are products of the isoprenoid biosynthetic pathway. Based on their chemical structure, the phytosterols can be grouped into 4-desmethyl, 4-monomethyl and 4,4-dimethyl sterols, being the most representative group in vegetable oils (Piironen et al., 2000; Martínez-Vidal et al., 2007).

Phytosterols are triterpenes similar to cholesterol both in structure and biological function. The basic skeleton of sterol molecule consists of 17 carbon atoms arranged in four rings; three six-membered rings (A, B and C) and one five-membered ring (D) (Figure 1).

The main olive oil sterols are β-sitosterol (75–90% of total sterols), Δ-5-avenasterol (5–20%), campesterol (1–4%) and stigmasterol (0.5–2%) (Figure 1). Other minor phytosterols that can be found in olive oil, with small amounts, are clerosterol, sitostanol, campestanol, cholesterol, 24-methylencholesterol, Δ-7-avenasterol, Δ-7-campesterol, Δ-5,24-stigmastadienol, Δ-7-stigmastenol and brassicasterol (Phillips et al., 2002, Mahjoub–Haddada et al., 2007; Lazzez et al., 2008; Manai-Djebali et al., 2012; Konuskan and Mungan, 2016).

For the Δ-5-avenasterol, larger concentrations up to 36%, have been reported for some olive oil cultivars (Boskou et al., 2006).

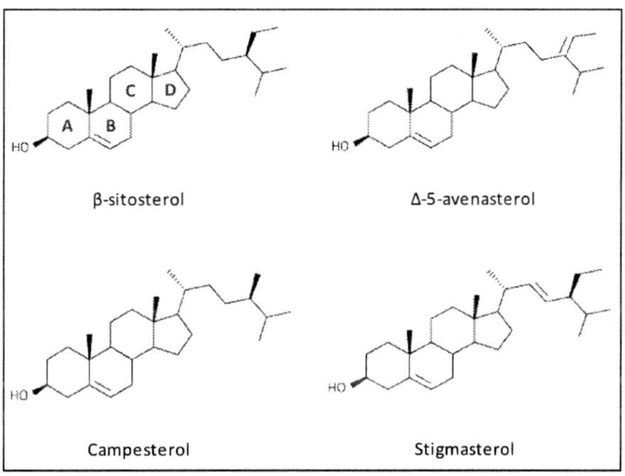

Figure 1. Structures of some representative phytosterols found in olive oil.

Sterols are important lipids related to the quality of olive oil. It have been reported that many olive oil phytosterols are associated with quality.

The concentration of Δ-5-avenasterol is important for olive oil quality, as the presence of an ethylidene group in this sterol confers exceptional antioxidant properties at high temperatures (Williamson et al., 1988;

Rossell, 2001). High amounts of Δ-5-avenasterol are associated with high oxidative stability (Manai-Djebali et al., 2017).

Stigmasterol is also related to various parameters of VOO quality. High levels correlate with high acidity and low organoleptic quality (Gracia, 2001; Noorali et al., 2014). Furthermore the ratio of campesterol/stigmasterol has been reported as a quality index of olive oil (Ranalli and Angerosa, 1996).

2. Main Factors Affecting Olive Oil Sterol Profile

Several researches have studied the influence of many factors on sterol profile. Olive oil sterol composition is affected by the variety, maturity stage of the olives, extraction method, environmental conditions and storage time prior to oil extraction (Fernandez-Cuesta et al., 2013; Yorulmaz et al., 2014; Konuskan and Mungan, 2016).

The total content of sterols and their composition in olive oil is strongly influenced by genetic factors (variety). Processing practices particularly affect stigmasterol while horticultural practices and fruit characteristics tend to affect more significantly other sterols such as β-sitosterol, sitostanol, Δ-5-avenasterol and Δ-7-avenasterol (Guillaume et al., 2012).

The horticultural and processing practices were: fruit size, maturity, malaxing time, malaxing temperature and delays between harvest and process.

2.1. Fruit Characteristics

Fruit characteristics such as maturity and size have a significant effect on olive oil sterol composition. Consequently, the relationships between them could potentially be used to determine optimal harvesting times.

The maturity index of olives was determined according to the method developed by the Agronomic Station of Jaén (Uceda and Hermoso, 1998) based on the evaluation of the olive skin and pulp colors. The maturity index values ranged from 0 (100% intensely green skin) to 7 (100% purple flesh and black skin).

Most of phytosterols are significantly affected by the maturation stage of the fruit, particularly β-sitosterol, apparent β-sitosterol, Δ-5-avenasterol, sitostanol, campesterol, sitostanol and Δ-7-avenasterol. Among them sitostanol is the most affected (Guillaume et al., 2012).

It was reported that campesterol and β-sitosterol decrease during ripening while Δ-5-avenasterol and Δ-7-avenasterol significantly increase (Guillaume et al., 2012). At early stages of maturation campesterol content is very close to 4.5%, which is above the upper limit set by the Official Norm (EU Regulation, 2016; IOC, 2016), and it decreases afterward (Koutsaftakis et al.,1999). Which would indicate that the oil sterol composition improved with olive maturation.

Other authors stated that the ratio campesterol/stigmasterol (as quality index) and Δ-5-avenasterol reached the highest values while β-sitosterol reached its minimum value when olives were harvested at their optimum maturation degree (Koutsaftakis et al., 1999).

Several studies reported that the total sterol amounts decline with ripening process, this is because the sterols became diluted when the oils are accumulated in the fruit during the maturation (Mariani et al., 1991; Gutiérrez et al., 1999; Boulkroune et al., 2017).

Moreover, it was mentioned that olive size affects significantly the phytosterol contents. Campestanol, stigmasterol, β-sitosterol, sitostanol, Δ-5-avenasterol and Δ-7-avenasterol are significantly affected by fruit size. While β-sitosterol, sitostanol significantly decrease with fruit size, Δ-5-avenasterol and Δ-7-avenasterol increase (Guillaume et al., 2012; Guillaume and Ravetti, 2015).

Along the ripening, several metabolic processes take place in olives with subsequent variations on the profiles of some compounds. Several studies conducted on oils extracted from olives at different maturity stages had characterized the modification of the sterol profile during the olive development. It has been proven that the ripening stage had a significant effect on the sterolic contents and the results showed a decreasing level of total sterols, β-sitosterol and an increasing level of Δ-5-avenasterol according to the maturity process (Gutiérrez et al., 1999; Aparicio and Luna, 2002; Vekiari et al., 2010; Boulkroune et al., 2017).

As Δ-5-avenasterol, is a precursor in the biosynthesis of β-sitosterol (Clouse, 2002), differential enzymatic efficiencies at different stages of fruit development might result in the accumulation of different levels of both sterols.

2.2. Variety

It is important to point out that variety is among the most important factors affecting the profile of the sterol fraction in olive oil (Lukić et al., 2013; Lazzez et al., 2008; Mailer et al., 2010; Matos et al., 2007; Vekiari et al., 2010).

Spanish (Rivera del Álamo et al., 2004; Sanchez-Casas et al., 2004; López-Cortés et al., 2013), Italian (Marini et al., 2004; Cercaci et al., 2007), Portuguese (Cunha et al., 2006; Matos et al., 2007), Tunisian (Stiti et al., 2002; Mahjoub–Haddada et al., 2007; Manai–Djebali et al., 2012), Turkish (Matthäus and Özcan, 2011, Yorulmaz et al., 2013; Konuskan and Mungan, 2016), Australian (Mailer et al., 2010; Guillaume et al., 2012), Greek (Vekiari et al., 2010) and Croatian (Lukić et al., 2013) olive varieties have been evaluated for their sterol composition. All these previous works have demonstrated that sterol profile and content was significantly affected by the olive oil variety.

The study of Guillaume and coworkers clearly demonstrates the strong influence of the variety on sterol composition, particularly in the case of certain sterols such as campesterol, stigmasterol, β-sitosterol and total sterols (Guillaume et al., 2012).

Several successful attempts have been made to characterize and differentiate olive oils according to varietal origin on the basis of sterols

(Bianchi et al., 2001; Ranalli et al., 2002; Galeano Diaz et al., 2005; Mahjoub–Haddada et al., 2007).

High abundance of β-sitosterol, and significant differences between its concentration ranges in oils from different varieties, presumably ensured its robustness as an indicator of varietal origin. Interestingly, similar results were obtained by several studies (Koutsaftakis et al., 1999; Mahjoub–Haddada et al., 2007; Manai–Djebali et al., 2012; López-Cortés et al., 2013; Lukić et al., 2013).

The most contributor sterols to classify olive oil samples according to the cultivar are: Δ-7-stigmastenol, 24-methylencholesterol, clerosterol, campesterol, Δ-7-campesterol, β-sitosterol and apparent β-sitosterol (Sanchez-Casas et al., 2004; Manai–Djebali et al., 2012; Lukić et al., 2013).

A new parameters, the 24-methylene-cholesterol/stigmasterol and Δ-7-campesterol/Δ-5,24-stigmastadienol ratios emerged as reliable variety indicators, these two ratios contributing significantly to the variety differentiation (Lukić et al., 2013).

2.3. Geographical Origin

Many findings reported that geographical region had a significant effect on sterol amounts this may be related to the different climate conditions in each growing area, such as altitude, rainfall, temperature and humidity (Ben Temime et al., 2008; Piravi-Vanak et al., 2012; Bajoub et al., 2015; Arafat et al., 2016).

It has been demonstrated that total sterols, β-sitosterol, Δ-5-avenasterol and stigmasterol are the most important variables for the characterization of Tunisian olive oils according to their geographical origin (Ben Temime et al., 2008; Ouni et al., 2011).

These results are in agreement with other research which showed that campesterol, stigmasterol, Δ-5-avenasterol, β-sitosterol, Δ-5,24-stigmastadional and total sterols contents were significantly affected by the growing region (Mailer et al., 2010; Konuskan and Mungan, 2016).

Additional studies reported that the concentration of β-sitosterol increased and those of campesterol and stigmasterol decreased as the climate became colder (Lotti et al., 1992; Piravi-Vanak et al., 2012).

Giacalone and coworkers (2015) highlighted the high significance of sterols to characterize extra virgin olive oils in relation to their countries origin. Multivariate statistical analysis using sterol data gave a perfect differentiation of Italian extra virgin olive oils from those of other countries.

Qualitative and quantitative analyses could be used to characterize extra virgin olive oil using sterols as markers of the growing area. Some cultivars of olive oil are recognized as being of higher quality because they derive from well defined geographical areas, command better prices, and generally are legally protected. Indeed, the aim of Protected Designations of Origin (PDO) is to add value to certain specific high-quality products from a particular origin.

The European Union regulations provide guidelines for maintaining the PDO of olive oil. This includes characterization of olive oil based on cultivar and geographical origin, as they are used as an indicator of authenticity and quality. Therefore, there is an economic basis for identifying characteristics that distinguish PDO of olive oils. This has been tackled by many research institutions and has led to the application of different methods to authenticate PDO. Fatty acids and sterol compositions have been evaluated as a means of determining the origin of Tuscan oils (Armanino et al., 1989), Fatty acids for Greek (Tsimidou and Karakostas, 1993) and Italian (Sacco et al., 2000) oils, and triacylglycerls for oils from around the world (Gigliotti et al., 1993).

2.4. Extraction Procedure

The extraction procedure has also a great importance given that it has a significant influence on the sterol fraction.

Comparative extraction trials were carried out between classical pressing and centrifugation systems. The findings highlighted that campesterol, stigmasterol, β-sitosterol, Δ-5-avenasterol, and Δ-7-avenasterol are significantly affected by the extraction system. Among them stigmasterol was found to be the most affected; it was high in oils obtained with the pressing system. The campesterol/stigmasterol ratio, which has been reported as quality index, was significantly high in oils extracted by centrifugation systems (Ranalli and Angerosa, 1996; Koutsaftakis et al., 1999).

This result agrees with that of other researchers, which found that oils from the classical pressing system had significantly higher values for β-sitosterol and stigmasterol than the oils from centrifugation systems (Ranalli and Serraiocco, 1996).

The malaxing time of the olive paste is a very important parameter of good manufacturing practice. Stigmasterol and Δ-7-stigmastenol are significantly affected. These components tend to increase with more malaxing time (Guillaume et al., 2012).

Similarly to malaxing time, processing temperature is another important parameter during the olive oil manufacturing process. Stigmasterols tend to increase with a high malaxing temperature. Additionally, the total level of sterols was significantly affected by this processing parameter showing increasing values at higher malaxing temperatures (Guillaume et al., 2012).

Comparison of the different kneading temperatures showed that the most affected sterols are stigmasterol and Δ-5-avenasterol. Stigmasterol is high in the oils kneaded at 45°C whereas Δ-5-avenasterol and the campesterol/stigmasterol ratio are high when the kneading temperature is 30°C, which leads to the conclusion that high-quality oils are extracted at low temperatures (Koutsaftakis et al.,1999).

Similar to the other processing parameters, the delay between harvest and processing significantly affected the percentage of stigmasterol. Stigmasterol levels increased with long days between harvesting and processing. Campestanol was also significantly affected; however it decreased with larger delays between harvesting and processing (Guillaume et al., 2012).

3. Utility of Phytosterols in Authenticity and Discrimination of Olive Oils

Sterol composition is important for assessing authenticity, detecting adulteration, and characterization purposes because it is specific for each oil. Olive oils command a premium price in the market, leading to great temptation to adulterate by blending them with other lower grade oils such as refined olive oils, vegetable and seed oils.

3.1. Detection of Olive Oils Authenticity by Determination of their Sterol Composition

The authenticity and traceability of olive oils has been the objective of numerous studies in the past few years. Generally, characterization of monovarietal olive oils requires the study of several compounds, including major components, such as fatty acids (FA), triacylglycerols (TAG), and the minor ones, such as phytosterols, polyphenols, tocopherols, tocotrienols and pigments.

The determinaion of the sterol composition is very important in the evaluation of the nutritional value as well as in the quality control of olive oils, since they can be used for checking its genuineness and detecting adulteration. Furthermore, sterols can be considered as real fingerprint of olive oil (Boskou, 1996; Lukić et al., 2013; Pardo et al., 2013).

To prevent olive oil adulteration several international organizations, such as the Codex Alimentarius of the FAO/WHO, the European Union (EU) and the International Olive Council (IOC) have established characteristic values for some of these compounds.

The International Olive Council (IOC, 2016) and the European Union (EU Regulation, 2016) impose limits or ranges for some sterol compounds based on the natural levels found in traditional olive oil (Table 1).

Sterol profiles outside these ranges could suggest that the olive oil is not genuine. A number of researches have found olive oils which naturally exceed the limits for sterols. This is particularly in the case of campesterol which is specified to be less than 4% of total sterols according to the standard limits (EU Regulation, 2016; IOC, 2016). Some cultivars are characherized by campesterol content exceeding this limit such as Arbequina, Corniche, Koroneiki, Cornicabra, Arauco and Barnea (Koutsaftakis et al.,1999; Sanchez-Casas et al., 2004; Rivera del Álamo et al., 2004).

Compositional analysis of the olive oil sterol fraction can be used to assess the degree of purity of the oil and the absence of cheaper oils. Numerous adulterants have been found in virgin olive oil, and they vary from refined olive oil, deodorized virgin olive oils, raw olive pomace oil and synthetic olive oil–glycerol mixtures to almost all seed oils (e.g., soybean, maize, cottonseed, rapeseed and sunflower).

For instance, olive oil mixed with soybean, rapeseed, or sunflower oil can be detected on the basis of the concentrations of campesterol and stigmasterol. Besides, brassicasterol is present in mustard and rapeseed oils but absent or found only in trace amounts in olive oils, and thus it is a good indicator of adulteration of olive oil (Kamm et al., 2001; Mathison and Holstege, 2013). Moreover, it was demonstrated that the additions of around 2% rapeseed oil were detected by determining the content of brassicasterol (≤0.1% of total sterols for olive oil vs. 12–13% for rapeseed oil). The addition of 10% soybean oil was detected by determining the concentration of campesterol (15–24% for soybean oil vs. ≤4% for olive oil) and stigmasterol (16–19% for soybean oil vs. less than campesterol for olive oil). Contamination of sunflower oil higher than 5% could be detected by the increase of campesterol (7–13% for sunflower oil), stigmasterol (8–11% for sunflower oil), and Δ-7-stigmastenol (7–13% for sunflower oil vs. ≤0.5% for olive oil) concentrations. Adulteration with 10% or higher of grapeseed oil would increase the concentrations of campesterol (9–14% for grapeseed oil) and stigmasterol (9–17% for grapeseed oil) (Aparicio and Aparicio-Ruiz, 2000). As stated above, the analysis of sterol profile is almost decisive in clarifying the contamination of olive oil with other cheaper ones. This determination also permits characterization of the olive oil category as extra virgin, virgin, refined, etc. (Table 1) (EU Regulation, 2016; IOC, 2016).

Table 1. Olive Oil Purity Characteristics Established by the EU and the IOC (EU Regulation, 2016; IOC, 2016)

Category	Sterol Composition (%)						Total sterols (mg/kg)
	Cholesterol	Brassicasterol	Campesterol	Stigmasterol	App. β–sitosterol*	Δ-7-stigmastenol	
1. Extra virgin olive oil	≤ 0.5	≤ 0.1	≤ 4.0	< Camp.	≥ 93.0	≤ 0.5	≥ 1 000
2. Virgin olive oil	≤ 0.5	≤ 0.1	≤ 4.0	< Camp.	≥ 93.0	≤ 0.5	≥ 1 000
3. Lampante olive oil	≤ 0.5	≤ 0.1	≤ 4.0	__	≥ 93.0	≤ 0.5	≥ 1 000
4. Refined olive oil	≤ 0.5	≤ 0.1	≤ 4.0	< Camp.	≥ 93.0	≤ 0.5	≥ 1 000
5. Olive oil composed of refined and virgin olive oils	≤ 0.5	≤ 0.1	≤ 4.0	< Camp.	≥ 93.0	≤ 0.5	≥ 1 000
6. Crude olive-pomace oil	≤ 0.5	≤ 0.2	≤ 4.0	__	≥ 93.0	≤ 0.5	≥ 2 500
7. Refined olive-pomace oil	≤ 0.5	≤ 0.2	≤ 4.0	< Camp.	≥ 93.0	≤ 0.5	≥ 1 800
8. Olive-pomace oil	≤ 0.5	≤ 0.2	≤ 4.0	< Camp.	≥ 93.0	≤ 0.5	≥ 1 600

*App. β-sitosterol= Δ-5,23-stigmastadienol+chlerosterol+β-sitosterol+sitostanol+Δ-5-avenasterol+Δ-5,24-stigmastadienol.

The separation of sterol classes could be extremely useful in the monitoring of the refining process, which decreases free and increases esterified sterol contents. Lampante olive oils can be differentiated from extra virgin olive oils by the low ratio of free campesterol and free stigmasterol (Verleyen et al., 2002; Cunha et al., 2006). Refined oils contain low levels of total sterols because the refining process gives rise to significant loss of sterols (up to 25%) (Mailer et al., 2010).

According to Aparicio and Aparicio-Ruiz (2000), the qualitative and quantitative differences in phytosterol composition among several vegetable oils make them a suitable parameter to verify the botanical origin of vegetable oils and, thus, to detect eventual contaminations.

Furthermore the work of Alves and colleagues (2005) demonstrates that it was possible to evaluate the authenticity of PDO olive oils on the basis of a model developed for the characterization of monovarietal olive oils according to their sterol composition.

3.2. Discrimination of Olive Oil Varieties Using their Sterol Composition

For discrimination of monovarietal olive oils, the chemical data frequently need to be further explored by using statistical tools. Nowadays, several examples of statistical analysis applied to olive oil compositional data are available in the literature.

Multivariate statistical analysis has been used to recognize the parameters that are able to discriminate the olive varieties. Among chemometric methods, principal component analysis (PCA), hierarchical cluster analysis (HCA), and linear discriminant analysis (LDA) have gained increasing importance in recent years owing to promising results from characterization studies.

Several researchers investigated the usefulness of chemical data obtained through olive oil characterization on the discrimination of genetic variety. Among the components used in this discrimination there are fatty acids, triacylglycerols, aliphatic alcohols, sterols, triterpene dialcohols and tocopherols.

Results of Yorulmaz and colleagues (2014) revealed that sterol composition was more evident than fatty acid and triacylglycerol profiles for the discrimination of olive varieties. Therefore, they can be used for proper labeling and marketing of varietal oils.

It is well-known that sterol composition can be used to identify adulteration of olive oil, and it has been suggested that it may be used to classify virgin olive oils according to their fruit variety (Cañabate-Díaz et al., 2007; Lazzez et al., 2008) and to discriminate between different vegetable oils. The study of Lerma-García and coworkers (2008) demonstated that sterol profiles were able to perfectly classify samples belonging to eight different botanical origins (hazelnut, sunflower, corn, olive, soybean, avocado, peanut and grapeseed).

The feasibility of classifying extra virgin olive oil (EVOO) according to their genetic variety using the sterol profiles has been demonstrated. EVOO mainly produced at La Comunitat Valenciana (Spain) can be correctly classified with a high reliability by the sequential application of two LDA models, a method proposed by Lerma-García and colleagues (2011).

This method could be helpful for controlling the genetic origin of the olives used to obtain the EVOO (Lerma-García et al., 2009). It was very interesting for olive oil industry; since it provides a fast technique to establish the genetic variety of EVOO. This method could be easily extended to the classification of EVOO obtained from other genetic varieties coming from different areas by constructing new LDA models on the basis of sterol profiles (Lerma-García et al., 2011).

The investigation of Lukić and coworkers (2013) demonstrated that sterols and triterpene dialcohols could be used as reliable indicators of variety and ripening degree among virgin olive oils from three Croatian varieties (Buža, Črna and Rosinjola). Multivariate statistical methods such as PCA and LDA provided efficient differentiation models based on sterols and triterpene dialcohols data.

Discriminant analysis proved to be effective in discriminating between olive varieties when it was applied to the sterol data set. Among all sterols the stigmasterol could allow the segregation between varieties (López-Cortés et al., 2013).

Matos and coworkers evaluated the usefulness of three chemical parameters (compositions on tocopherols, sterols and fatty acids) as a tool to discriminate three Portuguese olive oil varieties (Cvs. Cobrançosa, Madural and Verdeal Transmontana). These authors found that the three cultivars were clearly discriminated and campesterol, stigmasterol, clerosterol, β-sitosterol, Δ-5-avenasterol, Δ-7-avenastenol are among the variables that contributed to the discrimination between olive varieties (Matos at al., 2007).

Boulkroune and colleagues (2017) characterized the sterolic and alcoholic fractions of four Algerian olive oil varieties (Chemlal, Mekki, Aghenfas and Buichret), they reported that the most discriminating variables were cholesterol, campestanol, Δ-7-stigmastenol, Δ-7-avenasterol, clerosterol, alcohol C22, alcohol C25, alcohol C27.

Ranalli and colleagues (2002) studied the contents of sterols, triterpene dialcohols and aliphatic alcohols in oils from seed (non-woody part of the kernel), pulp (mesocarp plus epicarp) and whole olive fruit from seven major Italian varieties. These authors found that the olive varieties or fruit oil kinds appeared to be effectively discriminated through the use of several multivariate methods applied to the data of their sterol or alcohol composition. In particular, PCA and HCA were very effective methods for discriminating the olive variety, whereas canonical discriminant analysis was a very effective technique for differentiating either the olive variety or the fruit oil kind.

Several researchers suggested the effectiveness of sterol composition on the classification of EVOO according to the variety. Some authors mentioned that statistical analysis applied to only sterol data are sufficient to discriminate EVOO according to their varieties, but others reported that these components are effective in olive variety discrimination when they are used together with other components such as fatty acids, triacylglycerols, aliphatic alcohols, triterpene dialcohols and tocopherols.

4. Main Health Benefits of Phytosterols

Several reports have demonstrated that phytosterols may protect against many chronic ailments including arteriosclerosis, ulcer, and inflammation (Berger et al., 2004).

In addition to their cholesterol-lowering actions, phytosterols have also been reported to possess anti-cancer effects. Animal studies of the anti-cancer function have shown that a diet containing 2% mixed phytosterols inhibits cancer cell growth, reduces tumor size, and decreases the incidence of colon, breast, and prostate cancers induced by specific carcinogens (Awad et al., 1995; Awad and Fink, 2000; Ju et al., 2004;Awad et al., 2001a).

4.1. Phytosterols and Cholesterol

One of the major factors involved in cardiovascular disease is high Low-Density Lipoprotein (LDL)-cholesterol concentration in the bloodstream. In contrast, High Density Lipoprotein (HDL)-cholesterol is linked to a reduced risk of coronary heart disease (Trichopoulou,1998; Yusuf et al., 2004).

Most investigations on phytosterols health benefits focus on the pharmaceutical approach of lowering LDL-cholesterol.

The health benefits derives from the fact that the sterols have a similar structure to cholesterol, and have the capacity to reduce intestinal cholesterol absorption (Plat and Mensink, 2005; Balk et al., 2006) and serum LDL-cholesterol concentrations (Piironen et al., 2000). Since the morbidity and mortality from cardiovascular disease have been dramatically reduced using cholesterol-lowering drugs (statins), the interest in plant sterols lies in their potential to act as a natural preventive dietary product (Piironen et al., 2000).

Phytosterols are absorbed only in trace amounts but inhibit the absorption of intestinal cholesterol including recirculating endogenous biliary cholesterol, a key step in cholesterol elimination (Ostlund, 2002). A dose of 2 g/day of phytosterols reduces LDL-cholesterol by 10% according to a recent meta-analysis of 41 studies (Katan et al., 2003), and little difference is observed between Δ-5-sterols and 5α-reduced sterols (stanols). Natural dietary intake varies from about 167–437 mg/day (Ostlund, 2002).

Sitostanol was shown to be more potent in reducing cholesterol absorption and serum cholesterol levels than β-sitosterol (Heinemann et al., 1986).

In general, the reduction in total and LDL-cholesterol increased with increasing daily doses of sterols up to a dose of 2g/day, beyond which no further cholesterol-lowering effect could be observed (Plat et al., 2000a). Thus, the National Cholesterol Education Program's Adult Treatment Panel III recommends for adults to consume 2 g of phytosterols daily to reduce LDL-cholesterol and cardiovascular disease risk (Expert Panel on Detection, 2001).

It has been reported that phytosterols consumed in moderate doses substantially enhanced fecal cholesterol excretion and reduced the percentage of intestinal cholesterol absorption relative to a phytosterol-deficient diet, and the effect was dose-dependent. The marked increase in biliary cholesterol excretion may represent an important mechanism by which phytosterols favorably alter lipid metabolism (Racette et al., 2010).

Phytosterols have been shown to reduce blood total cholesterol, and LDL-cholesterol levels in normocholesterolemic and hypercholesterolemic subjects without affecting levels of HDL-cholesterol (Plat et al., 2000b; Miettinen and Gylling, 2006; Abumweis et al., 2008).

Natural phytosterols are bioactive and have prominent effects on whole body cholesterol metabolism, cholesterol absorption efficiency, cholesterol biosynthesis and cholesterol excretion (Racette et al., 2015).

In addition to their cholesterol lowering properties, plant sterols possess anti-cancer, anti-inflammatory, anti-atherogenicity, and anti-oxidation activities, and should thus be of clinical importance, even for those individuals without elevated LDL-cholesterol (Berger et al., 2004).

4.2. Phytosterols and Inflammation

Further observations from both animal and human studies have demonstrated the anti-inflammatory effects of phytosterols (Bouic et al., 1999; Nashed et al., 2005; Devaraj et al., 2006; Jones et al., 2007).

Most of the investigations reported reductions in the levels of pro-inflammatory cytokines, including C-reactive protein, after consumption of dietary plant sterols (Othman and Moghadasian, 2011).

It has previously been shown that diets enriched with phytosterols (2%, w/w) contributed to improved lipid profiles and reduced atherosclerotic lesions in apolipoprotein E-knockout (apo E-KO) mice (Nashed et al., 2005; Moghadasian, 2006).

These changes in lipids and atherosclerotic lesions have also been shown to be associated with reduced levels of pro-inflammatory cytokines, including IL-6 and TNF-a, and increased levels of anti-inflammatory IL-10 in spleen cells from lipopolysaccharide (LPS)-stimulated apo E-KO mice fed with phytosterols (Nashed et al., 2005).

Limited data from cell culture and animal studies suggest that phytosterols may attenuate the inflammatory activity of immune cells, including macrophages and neutrophils (Awad et al., 2004).

Among several non-lipid-lowering benefits suggested for phytosterols, anti-inflammatory effects were especially interesting, as atherosclerosis is known as an inflammatory disease. Thus, a combination of LDL-cholesterol lowering and anti-inflammatory effects of dietary phytosterols may provide protection against coronary artery disease (Othman and Moghadasian, 2011).

4.3. Phytosterols and Cancers

Apoptosis is a genetically programmed cell death that is essential for the development, the maintenance of tissue homeostasis, and the elimination of unwanted or damaged cells from multicellular organisms (Evans, 1993).

Unlike normal cells, cancerous cells lose their ability to respond to death signals that initiate apoptosis. Sitosterols have been found to induce apoptosis when added to cultured human prostate, breast and colon cancer cells (Von Holtz et al., 1998; Awad et al., 2003; Choi et al., 2003; Woyengo et al., 2009). Therefore, they may play significant roles in the management and prevention of human cancers.

Phytosterols have gained more insight for their protective effect against various forms of cancer. Phytosterols have been reported to be beneficial in reducing cancers of breast (Ju et al., 2004; Awad et al., 2007; Grattan, 2013), prostate (Awad et al., 2001b), lung (Mendilaharsu et al., 1998), liver, stomach (De Stefani et al., 2000; Woyengo et al., 2009; Ramprasath and Awad, 2015), ovary (McCann et al., 2003), and colon (Baskar et al., 2010).

β-sitosterol is the main phytosterol found in VOO and has been shown to inhibit proliferation and induce apoptosis in human solid tumors such as colon and breast cancers. Various findings have investigated the anticancer effect of this sterol.

The study of Park and colleagues (2007) demonstrated that β-sitosterol induced cytotoxicity and apoptosis in human leukemic U937 cells in a concentration dependent manner. Von Holtz and coworkers have shown that β-sitosterol inhibits the growth and stimulates apoptosis of prostate cancer LNCaP cells (Von Holtz et al., 1998).

Further works strongly suggest that β-sitosterol might be an effective natural component in cancer chemoprevention. Awad and coworkers (2001b; 2005) reported that β-sitosterol arrested human prostate and breast cancer cells in the G2/M phase of the cell cycle and induced apoptosis. These results suggest that the growth inhibitory and apoptosis inducing effects of β-sitosterol are the result of a block during this G2/M phase and such cells do not enter the G1 phase.

Moreover, β-sitosterol has been evaluated against a variety of cancers using preclinical experimental models (Park et al., 2007, Baskar et al., 2010). The studies have suggested potential efficacy of β-sitosterol particularly towards colon cancer, prostate cancer and leukemia (Awad et al., 1995; Von Holtz et al., 1998; Awad et al., 2007; Paniagua-Pérez et al., 2008).

A recent work of Manral and colleagues (2016) studies the effect of β-sitosterol on Pgp expression and cellular lipids in methyl nitrosourea (MNU) induced mammary gland carcinoma in female albino rats. These authors found that β-sitosterol can impart significant protection against the MNU induced mammary gland carcinogenesis through modifying the pathological, biochemical and inflammatory markers.

Reductions in the growth of various cancer cells including liver, prostate and breast by phytosterols treatment have been demonstrated. There have been several mechanisms proposed such as inhibition of carcinogen production, cancer cell growth and multiplication, invasion and metastasis and induction of cell cycle arrest and apoptosis. Other mechanisms including reduction of angiogenesis, invasion and adhesion of cancer cells and production of reactive oxygen species have also been suggested (Ramprasath and Awad, 2015).

4.4. Phytosterols and Diabetes

Various researches reported that phytosterol, especially β-sitosterol, have anti-diabetic and antioxidant effects in animal models. Radika and colleagues (2013) investigated the effects of β-sitosterol on insulin sensitivity, oxidative and nitrosative stress and lipid abnormalities in liver of high fat-fed rat model of insulin resistance. These authors found that β-sitosterol administration could promote sensitivity to insulin may be trough increasing nitric oxide levels in high fat diet rats (Radika et al., 2013).

Insulin resistance is a well recognized pathogenic factor that contributes to major health issues such as type 2 diabetes mellitus and cardiovascular disease (Reaven, 1995).

Other studies indicated that β-sitosterol treatment of diabetic rats prevented the development of diabetes. β-sitosterol reduced levels of glucose, nitric oxide and HbA1c in stereptozocin induced diabetic rats followed by increase in insulin level (Gupta et al., 2011). It also showed protective effect on pancreatic tissue with enhancement of pancreatic antioxidant (Vivancos and Juan, 2005; Gupta et al., 2011).

Thus, β-sitosterol has promising anti-diabetic as well as antioxidant effects and may be considered in clinical studies for drug development or as an adjuvant in the management of insulin resistance and associated fatty liver disease (Radika et al., 2013).

5. Conclusion

Phytosterols have a long history of consumption as food or pharmaceutical products, and generally recognized as safe without undesirable side effects. There are scientific evidences to support the fact that phytosterols have several biological activities which promote the health of animals and humans with only few adverse effects, such as occur in phytosterolemia, a rare genetic disorder.

These health benefits include reduction of LDL-cholesterol levels, which decrease the risk of cardiovascular diseases, prevention or treatment of various forms of cancers, anti-diabetic and anti-inflammatory activities.

For all these reasons, data concerning phytosterol composition are of high relevance for olive oil quality and health benefits and will enhance our knowledge and encourage olive oil consumption throughout the world.

References

Abumweis, S. S., Barake, R. & Jones, P. J. (2008). Plant sterols/stanols as cholesterol lowering agents: a meta-analysis of randomized controlled trials. *Food Nutr. Res.*,*52*, 1–17.

Alves, M. R., Cunha, S. C., Amaral, J. S., Pereira, J. A. & Oliveira, M. B. (2005). Classification of PDO olive oils on the basis of their sterol composition by multivariate analysis. *Anal. Chim. Acta*, *549*, 166–178.

Aparicio, R. & Aparicio-Ruiz, R. (2000). Authentication of vegetable oils by chromatographic techniques. *J. Chromatogr. A*, *881*, 93–104.

Aparicio, R. & Luna, G. (2002). Characterization of monovarietal virgin olive oils. *Eur. J. Lipid Sci. Technol.*, *104*, 614–627.

Arafat, S. M., Basuniy, A. M. M., Elsayed, M. E. & Soliman, H. M. (2016). Effect of pedological, cultivar and climatic condition on sterols and quality indices of olive oil. *Scientia Agric.*, *13*(1), 23–29.

Armanino, C., Leardi, R. & Lanteri, S. (1989). Chemometric Analysis of Tuscan Olive Oils, Chemom. *Intell. Lab. Sys.*, *5*, 343–354.

Awad, A. B., Chen, Y., Fink, C. S. & Hennessey, T. (1995). β-Sitosterol inhibits HT-29 human colon cancer cell growth and alters membrane lipids. *Anticancer Res.*, *16*(5A), 2797–2804.

Awad, A. B. & Fink, C. S. (2000). Phytosterols as anticancer dietary component: evidence and mechanism of action. *J. Nutr.*,*130*, 2127–2130.

Awad, A. B., Fink, C. S., Williams, H. & Kim, U. (2001a). *In vitro* and *in vivo* (SCID mice) effects of phytosterols on the growth and dissemination of human prostate cancer PC-3 cells. *Eur. J. Cancer Prev.*,*10*(6), 507–513.

Awad, A. B., Williams, H. & Fink, C. S. (2001b). Phytosterols reduce *in vitro* metastatic ability of MDA-MB-231 human breast cancer cells. *Nutr. Cancer*,*40*, 157–164.

Awad, A. B., Roy, R. & Fink, C. S. (2003). Beta-sitosterol, a plant sterol, induces apoptosis and activates key caspases in MDA-MB-231 human breast cancer cells. *Oncol. Rep.*,*10*(2), 497–500.

Awad, A. B., Toczek, J. & Fink, C. S. (2004). Phytosterols decrease prostaglandin release in cultured P388D1/MAB macrophages. *Prostaglandins Leukot. Essent. Fatty Acids*,*70*(6), 511–520.

Awad A. B., Burr A. T. & Fink C. S. (2005). Effect of resveratrol and β-sitosterol in combination on reactive oxygen species and prostaglandin release by PC-3 cells. Prostaglandins Leukot. *Essent. Fatty Acids*,*72*, 219–226.

Awad, A. B., Chinnam, M., Fink, C. S. & Bradford, P. (2007). β-Sitosterol activates Fast signaling in human breast cancer cells. *Phytomedicine*,*14*(11), 747–754.

Bajoub, A., Hurtado-Fernández, E., Ajal, E. A., Fernández-Gutiérrez, A., Carrasco-Pancorbo, A. & Ouazzani, N. (2015). Quality and chemical profiles of monovarietal north Moroccan olive oils from "Picholine Marocaine" cultivar: registration database development and geographical discrimination. *Food Chem.*,*179*, 127–136.

Balk, E. M., Lichtenstein, A. H., Chung, M., Kupelnick, B., Chew, P. & Lau, J. (2006). Effects of omega-3 fatty acids on serum markers of cardiovascular disease risk: a systematic review. *Atherosclerosis*,*189*, 19–30.

Baskar, A. A., Ignacimuthu, S., Paulraj, G. M. & Al Numair, K. S. (2010) Chemopreventive potential of β-Sitosterol in experimental colon cancer model - an *In vitro* and *In vivo* study. *BMC Complement. Altern. Med.*, *10*(24), 10p.

Ben Temime, S., Manai, H., Methenni, K., Baccouri, B., Abaza, L., Daoud, D., Sanchez-Casas, J., Osorio-Bueno, E. & Zarrouk, M. (2008). Sterolic composition of Chetoui virgin olive oil: influence of geographical origin. *Food Chem.*,*110*, 368–374.

Berger, A., Jones, P.J.H. & Abumweis, S. S.(2004). Plant sterols: factors affecting their efficacy and safety as functional food ingredients. *Lipids Health Dis.*,*3*, 1–19.

Bianchi, G., Giansante, L., Shaw, A. & Kell, D. B. (2001). Chemometric criteria for the characterization of Italian protected denomination of origin (DOP) olive oils from their metabolic profiles. *Eur. J. Lipid Sci. Technol.*,*103*, 141–150.

Boskou, D. (1996). *Olive oil chemistry and technology*. AOCS press, Champaign, pp. 12–52.

Boskou, D., Blekas, G. & Tsimidou, M. (2006). Olive oil composition. In D. Boskou (Ed.), Olive oil. *Chemistry and technology* (pp. 41–72). Champaign, IL: AOCS Press.

Bouic, P.J., Clark, A., Lamprecht, J., Freestone, M., Pool, E.J., Liebenberg, R.W., Kotze, D. & van Jaarsveld, P.P. (1999). The effects of β-sitosterol (BSS) and β-sitosterol glucoside (BSSG) mixture on selected immune parameters of marathon runners: inhibition of post marathon immune suppression and inflammation. *Int. J. Sports Med.,20*, 258–262.

Boulkroune, H., Lazzez, A., Guissous, M., Bellik, Y. Smaoui, S., Grati Kamoun N. & Madani, T. (2017). Characterization of sterolic and alcoholic fractions of some Algerian olive oils according to the variety and ripening stage. *OCL-Ol Corps Gras Li.* (DOI: 10.1051/ocl/2017026).

Cañabate-Díaz,B., Segura Carretero, A., Fernández-Gutiérrez, A., Belmonte Vega, A., Garrido Frenich, A., Martínez Vidal, J. L. & Duran Martos, J. (2007). Separation and determination of sterols in olive oil by HPLC-MS. *Food Chem.,102* (3), 593–598.

Cercaci, L., Passalacqua, G., Poerio, A., Rodriguez-Estrada, M.T. & Lercker, G. (2007). Composition of total sterols (4-desmethyl-sterols) in extra virgin olive oils obtained with different extraction technologies and their influence on the oil oxidative stability. *Food Chem.,102*, 66–76.

Choi, Y.H., Kong, K.R., Kim, Y.A., Jung, K.O., Kil, J.H., Rhee, S.H. & Park, K.Y. (2003).Induction of Bax and activation of caspases during beta-sitosterol-mediated apoptosis in human colon cancer cells. *Int. J. Oncol.,23*(6), 1657–1662.

Clouse, S. D. (2002). Arabidopsis mutants reveal multiple roles for sterols in plant development. *The Plant Cell,14*, 1995–2000.

Cunha, S.S., Fernandes, J.O. & Oliveira, M.B.P.P. (2006). Quantification of free and esterified sterols in Portuguese olive oils by solidphase extraction and gas chromatography–mass spectrometry. *J. Chromatogr. A,1128*, 220–227.

De Stefani, E., Boffetta, P., Ronco, A.L., Brennan, P., Deneo-Pellegrini, H., Carzoglio, J.C. & Mendilaharsu, M. (2000).Plant Sterols and Risk of Stomach Cancer: A Case-Control Study in Uruguay. *Nutr. Cancer,37*, 140–144.

Devaraj, S., Autret, B.C. & Jialal, I. (2006). Reduced-calorie orange juice beverage with plant sterols lowers C-reactive protein concentrations and improves the lipid profile in human volunteers. *Am. J. Clin. Nutr.,84*, 756–761.

EFSA (2008). Plant Sterols and Blood Cholesterol. *EFSA J.,781*, 1–12.

EU Regulation (2016). Commission Delegated Regulation EU 2016/2095 of 26 September 2016 amending Regulation (EEC) No 2568/91 on the characteristics of olive oil and olive-residue oil and on the relevant methods of analysis. *Off.J. Eur. Union L,326*, 1–6.

Evans, V.G. (1993). Multiple pathways to apoptosis. *Cell Biol. Int.,17*(5), 461–476.

Expert Panel on Detection (2001). Executive summary of the third report of the national cholesterol education program (NCEP) Expert Panel on Detection, Evaluation, and Treatment of High Blood Cholesterol in Adults (Adult Treatment Panel III). *JAMA,285*, 2486–2497.

FDA (2010). Department of Health and Human Services. Food Labeling; Health Claim; Phytosterols and Risk of Coronary Heart Disease; Proposed Rule. *Federal Register,75*, 76526-76571.

Fernandez-Cuesta, A., Leon, L., Velasco, L. & De La Rosa, R. (2013). Changes in squalene and sterols associated with olive maturation. *Food Res. Int.,54*, 1885–1889.

Galeano Diaz, T., Durán Merás, I., Sánchez Casas, J. & Alexandre Franco, M.F. (2005). Characterization of virgin olive oils according to its triglycerides and sterols composition by chemometric methods. *Food Control,16*, 339–347.

Giacalone, R., Giuliano, S., Gulotta, E., Monfreda, M. & Presti, G. (2015). Origin assessment of EV olive oils by esterified sterols analysis. *Food Chem.,188*, 279-285.

Gigliotti, C., Daghetta A. & Sidoli A. (1993). Indagine Conoscitiva sul Contenuto Trigliceridico di Oli Extra Vergini di Oliva di Varia Provenienza. *Riv. Ital. Sostanze Grasse,70*, 483–489.

Gracia, M.S. (2001). Composicion quimica de distintas calidades de aceites de oliva virgen de la variedad «Empetre» en el bajo Aragon. *Grasas Aceites,52*, 52–58.

Grattan, B.J.Jr. (2013). Plant sterols as anticancer nutrients: evidence for their role in breast cancer. *Nutrients,5*, 359–387.

Guillaume, C., Ravetti, L., Ray, D. L. & Johnson, J. (2012). Technological factors affecting sterols in Australian olive oils. *J. Am. Oil Chem. Soc.,89*, 29–39.

Guillaume, C. & Ravetti, L. (2015). Technological and agronomical factors affecting sterols in Australian olive oils. *Riv. Ital. Sostanze Grasse,92* (1), 53-60.

Gupta, A., Sharma, A.K., Dobhal, M.P., Sharma, M.C. & Gupta, R.S. (2011). Antidiabetic and antioxidant potential of β-sitosterol in treptozotocin-induced experimental hyperglycemia. *J. Diabetes,3*, 29–37.

Gutiérrez, F., Jiménez, B., Ruíz, A. & Albi, M.A. (1999). Effect of olive ripeness on the oxidative stability of virgin olive oil extracted from the varieties Picual and Hojiblanca and on the different components involved. *J. Agric. Food Chem.,47*, 121–127.

Gutiérrez, F., Varona, I. & Albi, M. A. (2000). Relation of acidity and sensory quality with sterol content of olive oil from stored fruit.*J. Agric. Food Chem.,48*, 1106–1110.

Heinemann, T., Leiss, O. & van Bergmann, K. (1986). Effect of low dose sitostanol on serum cholesterol in patients with hypercholesterolemia. *Atherosclerosis,61*, 219–223.

IOC (2016).Trade standard applying to olive oils and olive pomace oils.COI/T.15/NC No 3/Rev. *11*. 2016, 4–5.

Jones, P.J., Demonty, I., Chan, Y.M., Herzog, Y. & Pelled, D.(2007). Fish-oil esters of plant sterols differ from vegetable-oil sterol esters in triglycerides lowering, carotenoid bioavailability and impact on plasminogen activator inhibitor-1 (PAI-1) concentrations in hypercholesterolemic subjects. *Lipids Health Dis.,6*, 1–9.

Ju, Y.H., Clausen, L.M., Allred, K.F., Almada, A.L. & Helferich, W.G. (2004). β-Sitosterol, β-Sitosterol Glucoside, and a Mixture of β-Sitosterol and β-Sitosterol Glucoside Modulate the Growth of Estrogen-Responsive Breast Cancer Cells *In Vitro* and in Ovariectomized Athymic Mice. *J. Nutr.,134*, 1145–1151.

Kamm, W., Dionisi, F., Hischenhuber, C.& Engel, K. (2001). Authenticity assessment of fats and oils. *Food Rev. Int.,71*, 249–290.

Katan, M.B., Grundy, S.M., Jones, P., Law, M., Miettinen, T. & Paoletti, R. (2003). Efficacy and safety of plant stanols and sterols in the management of blood cholesterol levels. *Mayo. Clin. Proc.*,*78*(8), 965–978.

Konuskan, D.B. & Mungan, B. (2016). Effects of variety, maturation and growing region on chemical properties, fatty acid and sterol compositions of virgin olive oils. *J. Am. Oil Chem. Soc.*,*93*, 1499–1508.

Koutsaftakis, A., Kotsifaki, F. & Stefanoudaki, E. (1999). Effect of extraction system, stage of ripeness, and kneading temperature on the sterol composition of virgin olive oils. *J. Am. Oil Chem. Soc.*,*76* (12), 1477–1481.

Lazzez, A., Perri, E., Caravita, M.A., Khlif, M. & Cossentini, M. (2008). Influence of olive maturity stage and geographical origin on some minor components in virgin olive oil of the Chemlali variety. *J. Agric. Food Chem.*,*56*, 982–988.

Lerma-García, M.J., Ramis-Ramos, G., Herrero-Martínez, J.M. & Simó-Alfonso, E.F. (2008). Classification of vegetable oils according to their botanical origin using sterol profiles established by direct infusion mass spectrometry. *Rapid Commun. Mass Spectrom.*,*22*, 973–978.

Lerma-García, M.J., Concha-Herrera, V., Herrero-Martínez, J.M. & Simó-Alfonso, E.F. (2009). Classification of Extra Virgin Olive Oils Produced at La Comunitat Valenciana According to Their Genetic Variety Using Sterol Profiles Established by High-Performance Liquid Chromatography with Mass Spectrometry Detection. *J. Agric. Food Chem.*,*57*, 10512-10517.

Lerma-García, M.J., Simó-Alfonso, E.F., Méndez, A., Lliberia, J.L. & Herrero-Martínez, J.M. (2011). Classification of extra virgin olive oils according to their genetic variety using linear discriminant analysis of sterol profiles established by ultra-performance liquid chromatography with mass spectrometry detection. *Food Res. Int.*,*44*, 103–108.

López-Cortés, I., Salazar-García, D.C., Velázquez-Martí, B. & Salazar, D.M. (2013). Chemical characterization of traditional varietal olive oils in East of Spain. *Food Res. Int.*,*54*, 1934–1940.

Lotti, G., Izzo, R. & Riu, R. (1992). Influence of climate on the fatty acids and sterol composition of olive oil. *Riv. Sci. Alimentazione,11*, 115–121.

Lukić, M., Lukić, I., Krapac, M., Sladonja, B. & Piližota, V. (2013). Sterols and triterpene diols in olive oil as indicators of variety and degree of ripening. *Food Chem.,136*, 251–258.

Mahjoub–Haddada, F., Manai, H., Oueslati, I., Daoud, D., Sánchez, J., Osorio, E. & Zarrouk, M. (2007). Fatty Acid, Triacylglycerol, and Phytosterol Composition in Six Tunisian Olive Varieties. *J. Agric. Food Chem.,55*, 10941–10946.

Mailer, R. J., Ayton, J. & Graham, K. (2010). The influence of growing region, cultivar, and harvest timing on the diversity of Australian olive oil. *J. Am. Oil Chem. Soc.,87*, 877–884.

Manai-Djebali, H., Krichene, D., Ouni, Y., Gallardo, L., Sánchez, J., Osorio, E., Daoud, D., Guido, F. & Zarrouk, M. (2012). Chemical profiles of five minor olive oils varieties grown in Central Tunisia. *J. Food Compos. Anal., 27*, 109–119.

Manai-Djebali, H., Oueslati, I., Martínez-Cañas, M.A., Zarrouk, M. & Sánchez-Casas, J. (2017). Improvement of the sterol and triacylglycerol compositions of Chemlali virgin olive oils through controlled crossing with Mediterranean cultivars. *J. Oleo Sci*. In Press.

Manral, C., Roy S., Singh, M., Gautam, S., Yadav, R.K., Rawat, J.K, Devi, U., Ansari, M.N., Saeedan, A.S. & Kaithwas, G. (2016). Effect of β-sitosterol against methyl nitrosourea-induced mammary gland carcinoma in albino rats. *BMC Complement. Altern. Med.,16*(260), 10p.

Mariani, C., Fedeli, E., Grob, K. & Artho, A. (1991). Indagine sulle variazioni dei componenti minori liberied esterificati di oli ottenuti da olive in funzione della maturazione e dello stoccaggio. *Riv. Ital. Sostanze Grasse,68*, 179–188.

Mariani, C., Bellan, G., Morchio, G. & Pellegrino, A. (1999). Free and esterified minor components of olive and hazelnut oils: their potential utilisation in checking oil blend. *Riv. Ital. Sost. Gras.,76*, 297–305.

Marini, F., Balestrieri, F., Bucci, R., Magrì, A.D., Magrì, A.L. & Marini, D. (2004). Supervised pattern recognition to authenticate Italian extra virgin olive oil varieties. *Chemom. Intell. Lab. Syst.*,*73*, 85–93.

Martínez-Vidal, J. L., Garrido-Frenich, A., Escobar-García, M. A. & Romero-González, R. (2007). LC-MS determination of sterols in olive oil. *Chromatographia*,*65*, 695−699.

Mathison, B. & Holstege, D. (2013). A rapid method to determine sterol, erythrodiol, and uvaol concentrations in olive oil. *J. Agric. Food Chem.*,*61*, 4506–4513.

Matos, L.C., Cunha, S.C., Amaral, J.S., Pereira, J.A., Andrade, P.B., Seabra, R.M. & Oliveira, B.P.P. (2007). Chemometric characterization of three varietal olive oils (Cvs. Cobrançosa, Madural and Verdeal Transmontana) extracted from olives with different maturation indices. *Food Chem.*,*102*, 406–414.

Matthäus, B. & Özcan, M.M. (2011). Determination of fatty acid, tocopherol, sterol contents and 1,2- and 1,3-diacylglycerols in four different virgin olive oil. *J. Food Process Technol.*,*2*, 1–4.

McCann, S.E., Freudenheim, J.L., Marshall, J.R. & Graham, S. (2003). Risk of Human Ovarian Cancer Is Related to Dietary Intake of Selected Nutrients, Phytochemicals and Food Groups. *J. Nutr.*,*133*, 1937–1942.

Mendilaharsu, M., De Stefani, E., Deneo–Pellegrini, H., Carzoglio, J. & Ronco, A. (1998). Phytosterols and risk of lung cancer: A case-control study in Uruguay. *Lung Cancer*,*21*, 37–45.

Miettinen, T.A. & Gylling, H. (2006). Plant stanol and sterol esters in prevention of cardiovascular diseases: a review. *Int. J. Clin. Pharmacol. Ther.*,*44*, 247–250.

Moghadasian, M.H. (2006). Dietary phytosterols reduce cyclosporine-induced hypercholesterolemia in apolipoprotein E-knockout mice. *Transplant.*,*81*, 207– 213.

Nashed, B., Yeganeh, B., HayGlass, K.T. & Moghadasian, M.H. (2005). Antiatherogenic effects of dietary plant sterols are associated with inhibition of proinflammatory cytokine production in Apo E-KO mice. *J. Nutr.*,*135*, 2438–2444.

Noorali, M., Barzegar, M. & Sahari, M.A. (2014). Sterol and fatty acid compositions of olive oil as an indicator of cultivar and growing area. *J. Am. Oil Chem. Soc.*,*91*, 1571–1581.

Ostlund, R.E.Jr. (2002). Phytosterols in human nutrition. *Annu. Rev. Nutr.*,*22*, 533–549.

Othman R.A. & Moghadasian, M.H. (2011). Beyond cholesterol-lowering effects of plant sterols: clinical and experimental evidence of anti-inflammatory properties. *Nutr. Rev.*,*69*(7), 371–382.

Ouni, Y., Flamini, G., Ben Youssef, N., Guerfel, M. & Zarrouk, M. (2011). Sterolic composition and triacylglycerols of Oueslati virgin olive oil: comparison among different geographic areas. *Inter. J. Food Sci. Tech.*,*46*, 1747–1754.

Paniagua-Pérez, R., Madrigal-Bujaidar, E., Reyes-Cadena, S., Alvarez-González, I., Sánchez-Chapul, L., Pérez-Gallaga, J., Hernández, N., Flores-Mondragón, G. & Velasco, O. (2008). Cell protection induced by β-sitosterol: inhibition of genotoxic damage, stimulation of lymphocyte production, and determination of its antioxidant capacity. *Arch. Toxicol.*,*82*(9), 615–622.

Pardo, J.E., Sena, E., Cuesta, M.A., Granell, J.D., Valiente, J. & Alvarez-Ort, M. (2013). Evaluation of potential and real quality of virgin olive oil from "Campos de Hellin" (Albacete, Spain). *J. Am. Oil. Chem. Soc.*,*90*, 851–862.

Park, C., Moon, D.O., Rhu, C.H., Choi, B.T., Lee, W.H., Kim, G.Y. & Choi, Y.H. (2007). β-Sitosterol Induces Anti-proliferation and Apoptosis in Human Leukemic U937 Cells through Activation of Caspase-3 and Induction of Bax/Bcl-2 Ratio. *Biol. Pharm. Bull.*,*30*(7), 1317–1323.

Phillips, K.M., Ruggio, D.M., Toivo, J.I., Swank, M.A. & Simpkins, A.H. (2002). Free and esterified sterol composition of edible oils and fats. *J. Food Compos. Anal.*,*15*, 123–142.

Piironen, V., Lindsay, D.G., A Miettinen, T., Toivo, J. & Lampi, A.M. (2000). Plant sterols: biosynthesis, biological function and their importance to human nutrition. *J. Sci. Food Agric.*,*80*, 939–966.

Piravi-Vanak, Z., Ghasemi, J.B., Ghavami, M., Ezzatpanah, H. & Zolfonoun, E. (2012). The Influence of Growing Region on Fatty Acids and Sterol Composition of Iranian Olive Oils by Unsupervised Clustering Methods. *J. Am. Oil Chem. Soc.,89*, 371–378.

Plat, J., Kerckhoffs, D.A. & Mensink, R.P. (2000a). Therapeutic potential of plant sterols and stanols. *Curr. Opin. Lipidol.,11*, 571–576.

Plat, J. & Mensink, R.P. (2005). Plant stanol and sterol esters in the control of blood cholesterol levels: Mechanism and safety aspects. *Am. J. Cardiol.,96*, 15–22.

Plat, J., van Onselen, E.N., van Heugten, M.M. & Mensink, R.P. (2000b). Effects on serum lipids, lipoproteins and fat soluble antioxidant concentrations of consumption frequency of margarines and shortenings enriched with plant stanol esters. *Eur. J. Clin. Nutr.,54*, 671–677.

Racette S.B., Lin X., Lefevre M., Spearie C.A., Most M.M., Ma L. & Ostlund, R.E.Jr. (2010). Dose effects of dietary phytosterols on cholesterol metabolism: a controlled feeding study. *Am. J. Clin. Nutr.,91*, 32–38.

Racette, S.B., Lin, X., Ma, L. & Ostlund, R.E.Jr. (2015). Natural Dietary Phytosterols. *J. AOAC Int.,98*, 679–684.

Radika, M.K., Viswanathan, P. & Anuradha, C.V. (2013). Nitric oxide mediates the insulin sensitizing effects of β-sitosterol in high fat diet-fed rats. *Nitric Oxide,32*, 43–53.

Ramprasath, V.R. & Awad, A.B. (2015). Role of phytosterols in cancer prevention and treatment.*J. AOAC Int.,98* (3), 1–4.

Ranalli, A. & Angerosa, F. (1996). Integral centrifuges for olive oil extraction. The qualitative characteristics of products, *J. Am. Oil Chem. Soc.,73*, 417–422.

Ranalli, A. & Serraiocco, A. (1996). Evaluation of characteristics of olive oil produced by innovative or traditional processing technologies. *Riv. Ital. Sostanze Grasse,73*, 303–315.

Ranalli, A., Pollastri, L., Contento, S., Di Loreto, G., Iannucci, E., Lucera, L. & Russi, F. (2002). Sterol and alcohol components of seeds, pulp, and whole olive fruit oils. Their use to characterize olive fruit variety by multivariates. *J. Sci. Food Agric.*,*82*, 854–859.

Reaven, G.M. (1995). Pathophysiology of insulin resistance in human disease. *Physiol. Rev.*,*75*, 473–486.

Rivera del Álamo, R.M., Fregapane, G., Aranda, F., Gomez-Alonso, S. & Salvador, M.D. (2004). Sterol and alcohol composition of Cornicabra virgin olive oil: the campesterol content exceeds the upper limit of 4% established by EU regulations. *Food Chem.*,*84*, 533–537.

Rossell, J.B. (2001). Factors affecting the quality of frying oils and fats. In: Frying: Improving Quality (ed. J.B. Rossell). CRC Press, Boca Raton, FL, chapter 7.

Sacco, A., Brescia, M.A., Liuzzi, V., Reniero, F., Guillou, C., Ghelli, S. & Van Der Meer, P.(2000). Characterization of Italian Olive Oils Based on Analytical and Nuclear Magnetic Resonance Determinations. *J. Am. Oil Chem. Soc.*,*77*, 619–625.

Saeidnia, S., Manayi, A., Gohari, A. R. & Abdollahi, M. (2014). The Story of Beta-sitosterol- A Review. *Eur. J. Med. Plants*,*4*, 590–609.

Salvador, M. D., Aranda, F.& Fregapane, G. (1998). Chemical composition of commercial Cornicabra virgin olive oil from 1995/96 and 1996/97 crops. *J. Am. Oil Chem. Soc.*,*75*, 1305–1311.

Sanchez-Casas, S., Osorio-Bueno, E., Montaño-García, A.M. & Martinez-Cano, M. (2004). Sterol and erythrodiol + uvaol content of virgin olive oils from cultivars of extremadura (Spain). *Food Chem.*,*87*, 225–230.

Stiti, N., Msallem, M., Triki, S. & Cherif, A. (2002). Etude de la fraction insaponifiable de l'huile d'olive de différentes variétés Tunisiennes. *Riv. Ital. Sostanze Grasse*,*79*(10), 357–363.

Trichopoulou, A., Lagiou, P. & Papas, A. (1998). Mediterranean diet: are antioxidants central to its benefits? *In Antioxidant Status, Diet, Nutrition and Health*, (A.M. Papas, ed.) pp. 107–118, CRC Press, Boca Raton, FL.

Tsimidou, M. & Karakostas, K.X. (1993). Geographical Characterization of Greek Virgin Olive Oil by Non-parametric Multivariate Evaluation of Fatty Acid Composition. *J. Sci. Food Agric.*,*62*, 253–257.

Uceda, M. & Hermoso, M. (1998). La calidad del aceite de oliva. In: Barranco, D., Fernàndez-Escobar, R., Rallo, L. (Eds.), El cultivo del olivo. Junta de Andalucía Ediciones Mundi-Prensa, Spain, pp. 547–572.

Vekiari, S. A., Oreopoulou, V., Kourkoutas, Y., Kamoun, N., Msallem, M., Psimouli, V.& Arapoglou, D. (2010). Characterization and seasonal variation of the quality of virgin olive of the Throumbolia and Koroneiki varieties from Southern Greece. *Grasas Aceites,61*, 221–231.

Velasco, J. & Dobarganes, C. (2002). Oxidative stability of virgin olive oil. *Eur. J. Lipid Sci. Technol.,104*, 661–676.

Verleyen, T., Forcades, M., Verhe, R., Dewettinck, K., Huyghebaert, A. & De Greyt, W. (2002). Analysis of free and esterified sterols in vegetable oils. *J. Am. Oil Chem. Soc.,79*, 117–122.

Vivancos, M. & Juan, J.M. (2005). β-Sitosterol modulates antioxidant enzyme response in RAW 264.7 macrophages. *Free Radic. Biol. Med.,39*, 91–97.

Von Holtz, R.L., Fink, C.S. & Awad, A.B. (1998). Beta-Sitosterol activates the sphingomyelin cycle and induces apoptosis in LNCaP human prostate cancer cells. *Nutr. Cancer,32*(1), 8–12.

Williamson, E.(1988). The antioxidant activity of Δ5-avenasterol. PhD Thesis, University of Reading, UK.

Woyengo, T., Ramprasath, V. & Jones, P. (2009). Anticancer effects of phytosterols. *Eur. J. Clin. Nutr.,63*(7), 813–820.

Xianga, C., Xua, Z., Liua, J., Lib, T., Yangc, Z. & Ding, C. (2017). Quality, composition, and antioxidant activity of virgin olive oil from introduced varieties at Liangshan. *LWT - Food Sci. Technol.,78*, 226–234.

Yorulmaz, A., Yavuz, H. & Tekin, A. (2014). Characterization of Turkish olive oils by triacylglycerol structures and sterol profiles. *J. Am. Oil Chem. Soc.,91*, 2077–2090.

Yusuf, S., Hawken, S., Ounpuu, S., Dans, T., Avezum, A., Lanas, F., McQueen, M., Budaj, A., Pais, P., Varigos, J. & Lisheng, L. (2004). Effects of potentially modifiable risk factors associated with myocardial infarction in 52 countries (the INTERHEART study):case-control study. *Lancet,364*, 937–952.

Zarrouk, W., Carrasco-Pancorbo, A., Zarrouk, M., Segura-Carretero, A. & Fernández-Gutiérrez, A. (2009). Multi-component analysis (sterols, tocopherols and triterpenic dialcohols) of the unsaponifiable fraction of vegetable oils by liquid chromatography–atmospheric pressure chemical ionization–ion trap mass spectrometry. *Talanta*, *80* (2), 15, 924–934.

In: Olive Oil
Editors: T. Fritjof and B. Henning
ISBN: 978-1-53612-563-4

Chapter 3

TUNISIAN OLIVE OIL: QUALITY, COMPOSITION AND ANTIOXIDANT PROPERTIES

Malika Drira[1], Ons Rekik[1] and Mohamed Bouaziz[1,2]
[1]Laboratory of Electrochemistry and Environment,
National Engineering School of Sfax (ENIS) University of Sfax,
Tunisia
[2]Higher Institute of Biotechnology of Sfax,
Departement of Food Technology. University of Sfax, Tunisia

ABSTRACT

Different monovarietal extra virgin olive oil cultivars in Tunisia (Chemlali, Chetoui, Oueslati, Neb Jmal, Toffehi, Fakharri, Gemri, Zalmati, Gemri-Dhokkar, Chemchali and Dhokkar) having different chemical composition, were investigated in terms of their quality indices, sterol content, phenolic composition and volatile composition, oxidative stability and sensory profile to show their classification.

The identification of phenolic compounds was based on separation by high performance liquid chromatography (HPLC) equipped with

Diode Array Detector (DAD) followed by chromatography-mass spectrometry analysis (LC-MS/MS).

However, the qualitative and quantitative differences in phenolic composition were observed among cultivars, thus identifying twenty phenolic compounds belonging to different phenolic types. Indeed, the sensory evaluation remains the gold standard for assessing the aroma. Olive oils are the only foods where sensory evaluation is a legal requirement. Oil samples are scored and classified according to three positive attributes: Fruity, Bitter and Pungent.

The antioxidant activity of olive oil extracts was evaluated by DPPH. Olive oil is rich in antioxidants (like vitamin E and poly phenols). Moreover, it is rich in mono unsaturated fatty acids unlike saturated fatty acids. The results indicate that the cultivar influence the quali - quantitative production of volatiles and consequently the final aroma of the oil.

Keywords: Tunisian cultivars, phenolic composition, volatile composition, oxidative stability, sensory profile, antioxidant activity

INTRODUCTION

Extra virgin olive oil is a staple food largely consumed in the Mediterranean Countries. Its benefits to health have prompted an increased demand worldwide (Luchetti, 2002). In Tunisia, olive oil production plays an important role in the agronomy and economy (Gharbi et al., 2015; Jabeur et al., 2015). Accounting for more than 4% of the world olive oil production, Tunisia holds an important position in the olive oil market; it exports about 75% of its production and is ranked as the second largest exporter after the European Union with an average of 115,000 tons per year over the last five years (ONH, 2015).

Since virgin olive oil (VOO) is obtained from healthy olive drupes using only mechanical processing, it is edible without further refinement. Besides, owing to its sensory properties and nutritional value deriving from its high monounsaturated/polyunsaturated fatty acid ratio (Kiritsakis., 1990) together with natural antioxidant content (Boskou., 2006) have led to the increase in its consumption throughout the world. The fine sensory features of this fruit juice having a unique aroma and taste are essentially attributed to the presence of minor components, mainly volatile and phenolic compounds (Angerosa et al., 2002; Aparicio et al., 2002). Volatiles are the main responsible for the aroma of virgin olive oil, especially for the green sensory notes of high-quality extra virgin olive oils. However, compounds with a phenolic structure have an impact on both taste, especially the positive bitterness organoleptic attribute, and the oxidative stability of virgin olive oil (Angerosa et al., 2000). Tunisian orchards are rich in many varieties that have been the subject of a catalogue listing fifty-six different varieties (Trigui et al., 2002). Two major varieties dominate the Tunisian olive grove, namely the Chetoui in the North and Chemlali in the Centre and the South. Among the most important factors associated with the quality evaluation of olive oil, the olive ripening stage can be mentioned (Criado et al., 2005; Baccouri et al., 2007). In fact, during the ripening stage, numerous metabolic processes occur in olives with ensuing variations on the chemical structure and concentration of some compounds. Such changes are reflected in the quality grade, sensorial characteristics, oxidative stability and nutritional value of the obtained product. Examples of compounds involved in this phenomenon are chlorophyll, carotenoid pigments, and polyphenol and fatty acid compositions. Nevertheless, the phenolic and volatile contents and profiles of virgin olive oils depend on both (I) the initial composition characteristics of the olive fruits employed for processing and (II) the technological factors used during the oil mill process. The chemical composition of olive oil rely on some agronomical factors such as olive cultivar (Ammar et al., 2014; Ben Mansour et al., 2015a; Ben Mansour et al., 2016), the ripening stage of the fruit (Ben Brahim et al., 2016; Ben Mansour et al., 2015a; Gómez-Rico et al., 2008) and pedoclimatic conditions (Ben Mansour et al., 2015b; Gargouri et al., 2013).

Chemical and analytical work to elucidate the structure and to quantify minor constituents of olive oil is now progressing rapidly. Much work has been also carried out by nutritionists on the key functional components. This chapter examines what makes olive oil chemically different from other culinary fats in Tunisia. The discussion does not cover olive oil chemistry and technology completely. It aims to highlight the issues related to compositional characteristics and properties that make this oil so distinct.

Materials and Methods

Analytical Methods

Quality Indices Determinations

The determination of peroxide value (PV) (meqO_2/kg of oil), free acidity (given as % of oleic acid), K_{232} and K_{270} were carried out following the analytical methods described by the International Olive Council (International Olive Council 2010).

Pigment Content

Carotenoids and chlorophylls (mg/kg of oil) were determined at 470 and 670 nm, respectively, in cyclohexane using the specific extinction values prescribed by Haddada et al. (2008).

$$\text{Chlorophyll (mg/kg)} = (A_{670} * 10^6) / (613 * 100 * d)$$

$$\text{Carotenoid (mg/kg)} = (A_{470} * 10^6) / (2.000 * 100 * d);$$

where A is the absorbance and d is the spectrophotometer cell thickness (1 cm). The chlorophyll and carotenoid concentrations are expressed as mg of pheophytin α and lutein per kg of oil, respectively.

Fatty Acid Methyl Ester Analysis

The fatty acid composition of the oils was determined by gas chromatography (GC) as fatty acid methyl esters (FAMEs). FAMEs were prepared by saponification/ methylation with sodium methylate according to the European Union Commission Regulation EEC 2568/91 (1991). A chromatographic analysis was performed in a SHIMADZU set 17 A Series

II gas chromatography using a capillary column (stabilwax, Restek, length 50 m, internal diameter 0.32 mm and film thickness 0.25 lm). The column temperature was isothermal at 180°C; the injector and detector temperatures were 230 and 250°C, respectively.

Generally, fatty acids are identified by comparing retention times with standard compounds, and in the present study ten fatty acids were considered: palmitic (16:0), hypogeic (16:1n-9), palmitoleic (16:1n-7), heptadecanoic (C17:0), heptadecenoic (C17:1n-8), stearic (18:0), (Z)-vaccenic (18:1n-7), oleic (18:1n-9), linoleic (18:2), linolenic (18:3), arachidic (20:0) and gondoı¨c (C20:1n-9) acids expressed as percentages of fatty acid methyl esters.

Sterol Determination

The analysis of sterolic fractions extracted from olive oil was determined according to the method adopted by the European Union Commission Regulation EEC 2568/91 (1991). Briefly, 5 g of olive oil from each variety was first added to α-cholestanol used as internal standard and then saponified with potassium hydroxide solution in ethanol (2 M). After 1 h of boiling, 100 ml of water was added and the extraction of unsaponifiable was carried out by 200 ml of ethyl ether. Afterwards, 20 mg of unsaponifiable fraction was dissolved in 0.5 ml of chloroform and then separated by Silica gel plate chromatography. The elution was achieved by a mixture of hexane and ether (65/35, v/v); then the plate was pulverized using a solution of 2,7 dichlorofluorescein (0.2% in the ethanol). The band corresponding to sterols was then scraped. Sterols recovered from the plate were dissolved in chloroform and filtered.

The solvent was evaporated under N2 and the sterols were transformed into trimethylsilyl ethers by the addition of pyridine-hexamethyldisilane-trimethylchlorosilane (9:3:1, v/v/v) (Supelco, Bellefonte, PA, USA), left for 15 min and then centrifuged. Subsequently, the mixture was analysed by gas chromatography using a Gas Chromatograph System (model 7890A, Agilent Technologies) equipped with a flame ionization detector (FID). The column used was a capillary HP-5 (5% phenyl:95% dimethylpolysiloxane, length 30 m, id 0.32 mm and film thicknes1s 0.25

lm), and the operating conditions for the analysis were as follows: The oven temperature was isothermal at 260°C, temperature of injector was 280°C and detector was set at 290°C, helium was the carrier gas, with a flow through the column of 1 mL/min and 1:50 Split ratio and the injection volume was 5 μL.

Volatile Compound Analyses

Extraction

Solid-phase micro extraction was used as a technique for headspace sampling of virgin olive oils. Supelco SPME devices coated with polydimethylsiloxane (PDMS, 100 μm) were used to sample the headspace of 2 mL of olive oil inserted into a 4 mL glass septum vial and allowed to equilibrate for 30 min at 22 ± 1°C. After the equilibration time, the fiber was exposed to the headspace for 50 min. Extraction temperature and time were optimized in previous studies (Campeol et al., 2001).

Identification

The GC analyses were performed using a HP-5890 series II instrument (Palo Alto, CA (U.S.A.)) equipped with a DB-5 (Agilent, Santa Clara, CA (U.S.A.)) capillary column (30 m *0.25 mm i.d., 0.25 μm film thickness), working with the following temperature program: 60°C for 10 min, ramp of 5°C min 1 to 220°C; injector and detector temperatures, 250°C; carrier gas, helium (1 mL min^{-1}); detector FID; splitless injection. The components were identified by the comparison of their retention time with that of pure authentic standards and by means of their linear retention indices (LRI) relative to the series of *n*-hydrocarbons. As for the GC–EIMS analyses, they were performed using a Varian CP 3800 gas-chromatograph equipped (Palo Alto, CA(U.S.A.)) with a DB-5 capillary column (30 m *0.25 mm i.d., 0.25 μm film thicknesses) and a Varian Saturn 2000 (Palo Alto, CA(U.S.A.)) ion trap mass detector. The analytical conditions were as follow: injector and transfer line temperature at 250 and 240°C, respectively; oven temperature was programmed from 60 to 240°C at 3°C min^{-1}; carrier gas, helium at 1 mL min^{-1}; splitless injection. The identification of the constituents was based on the comparison of the retention time with that of the authentic samples, comparing their LRI relative to the series of *n*-alkanes, and on computer matching against commercial (NIST 98 and ADAMS) and home-made library mass spectra built from pure substances and the components of known oils and MS literature data (Adams., 1995). Moreover, the molecular weights of all the identified substances were confirmed by GC–CIMS, using MeOH as CI ionizing gas (Flamini et al., 2003).

Analysis of Phenolic Compounds

Extraction of Phenolic Fraction from Olive Oil

The phenolic extracts were obtained following the procedure of Chtourou et al., 2013, with some modifications. In fact, the oil sample (4g) was added to 2 mL of *n*-hexane and 4 mL of a methanol/water (60:40, v/v) mixture in a 20 mL centrifuge tube. After vigorous mixing, they were centrifuged for 3 min. The hydroalcoholic phase was collected, and the hexane phase was re-extracted twice with 4 mL of the methanol/water (60:40, v/v) solution each time. Finally, the hydroalcoholic fractions were combined, washed with 4 mL of *n*-hexane to remove the residual oil, then concentrated and dried by evaporative centrifuge in vacuum at 35°C.

HPLC-DAD Analysis of Phenolic Compounds

The apparatus used was a Shimadzu system consisting of a LC-10ATvp pump and a SPD-10Avp UV–visible detector. The stationary phase was a VP-ODS column (Shimadzu) maintained at 40°C. The mobile phase was 0.1% phosphoric acid in water (A) and 70% acetonitrile in water (B) at a constant flow rate of 0.6 mL min^{-1}. The solvent gradient started with 20% B and increased to 50% B in 30 min, followed by an isocratic plateau for 5 min. Finally, solvent B decreased to 20% until the end of the running time. With respect to the detection and quantification, they were performed at 280 nm. Peaks were identified by congruent retention times compared with standards. Each phenolic compound was quantified in comparison with its standard when it was available. Due to the absence of standards for some detected compounds, hydroxytyrosol glucoside was expressed as mg hydroxytyrosol/kg of oil; secoiridoids and lignans were expressed as mg oleuropein/ kg of oil.

Determination of Total Phenols

The determination of the total phenolic compounds included the use of the Folin–Ciocalteu reagent and the method was adopted from Bouaziz et al., (2010). The total phenolic content was expressed as milligrammes of gallic acid (GA) equivalents per kilogramme of oil (y = 0.0012x-0.0345, $R^2 = 0.9997$).

Determination of Total Flavonoids

Total flavonoids were measured by a colorimetric assay developed by Bouaziz et al., (2010); 1 mL aliquot of appropriately diluted sample or standard solutions of catechin (20, 40, 60, 80 and 100 mg L^{-1}) was added to a 10 mL volumetric flask containing 4 mL double-distillate.

H_2O. At zero time, 0.30 mL 5% $NaNO_2$ was added to the flask. After 5 min, 0.30 mL 10% $AlCl_3$ was added. At 6 min, 2 mL (1 mol L^{-1}) NaOH was added to the mixture.

Immediately, the reaction flask was diluted to volume with the addition of 2.40 mL of double-distillate H_2O and thoroughly mixed. Concerning the absorbance of the mixture, pink in colour, it was determined at 510 nm versus prepared water blank. As for the total flavonoids of fruits, they were expressed on a fresh weight basis as mg 100 g^{-1} catechin equivalents (CE). It is worth noting that the samples were analysed in triplicate.

Antioxidant Activity

Free Radical Scavenging Activity

The capacity to scavenge the free radical 2,2-diphenyl-1-picrylhydrazyl (DPPH) was monitored following the method of Bouaziz et al. (2010). The extract solution (0.25 mL) was mixed with 0.5 mL of methanolic solution containing DPPH radicals ($6 * 10^{-6}$ M). The mixture was shaken vigorously and left to stand for 30 min at room temperature in dark (until stable absorbance values were obtained). The reduction of the DPPH-radical was measured by continuous monitoring of the absorption decrease at 517 nm.

DPPH scavenging effect was calculated as the percentage of DPPH discoloration using the following equation:

$$\% \text{ scavenging effect} = [(A_{DPPH} - A_E)/A_{DPPH}] * 100;$$

Where A_E is the absorbance of the solution when the sample extract has been added at a particular level, and A_{DPPH} is the absorbance of the DPPH solution. Regarding the extract concentration providing 50% inhibition (IC_{50}), it was calculated from the graph of scavenging effect percentage against extract concentration in the solution.

Tocopherol Quantification

For the determination of α-, β- and γ-tocopherols, a solution of 250 mg oil in 25 mL of *n* heptane was used for HPLC. The HPLC analysis was conducted using a Merck–Hitachi low pressure gradient system, fitted with an L-6000 pump (Merck–Hitachi, Darmstadt, Germany).

a Merck–Hitachi F-1000 Fluorescence Spectrophotometer (Merck–Hitachi) with detector wavelengths for excitation 290 nm and a ChemStation integration system (Agilent Technologies Deutschland GmbH, Böblingen, Germany). Twenty microliters of the sample were injected by a Merck 655-A40 Autosampler (Merck-Hitachi, Darmstadt, Germany) onto a LiChrospher 100Å diol-phase HPLC column 25 cm × 4.6 mm ID (Merck, Darmstadt, Germany) used with a flow rate of 1.3 mL/min at 70 bar pressure. The used mobile phase was *n*-heptane/*tert*-butyl methyl ether (99:1, v/v). Identification of alpha-tocopherol was performed by comparing the retention time to a pure standard; a calibration curve was constructed with the standard solution of α-tocopherol (R^2 = 0.998) and used for quantification. A pure α-tocopherol was used as the external standard for the calibration curve, while α-tocopherol, β-tocopherol and γ-tocopherol were used as pure reference compounds for the identification. Results were given as milligrams tocopherols per kilogram oil.

Determination of Oxidative Stability by Rancimat Apparatus

The oil samples (5.0 g) were heated in Rancimat equipment at 120°C (Metrohm Ltd., Herisau, Switzerland). Oxidative stability was carried out as previously described (Chtourou et al., 2013).

Sensory Assessment

Sensory evaluation was determined by a panel of eight judges (males between 45 and 52 years) that were wholly qualified in the VOO assessment according to the official methods of the International Olive Oil Council (Conseil Oléicole International, 2012).

Results and Discussion

Quality Indices

The quality indices of Chemlali, Oueslati and Neb Jmel are listed in Table 1. The free acidity is not only an important quality factor, but also has been extensively used as a traditional criterion for classifying olive oil in various commercial grades (International Olive Council 2010). The free fatty acid content of all analyzed samples did not exceed the upper limit of 0.8% established by the COI norm and they corresponded to the extra VOO class (Table 1). The peroxide values evaluate the hydroperoxide content and offer a measure of lipid oxidation. In the samples study, these values range from 7.35 ± 0.19 to 13.81 ± 3.53 meqO_2/kg oil. With respect to the values of K_{232} and K_{270}, they vary between 1.54 ± 0.06 - 2.20 ± 0.00 and 0.12 ± 0.00 - 0.17 ± 0.01, respectively. As for the values of the analytical parameters, they range in the levels established for the highest quality category "extra virgin" olive oil European Community, Commission Regulation 702/2007 of 21 June (2007) and show that the geographical origin has significant influence (Table 1).

The contents in Chlorophylls and carotenoids in olive oils of Chemlali, Ouselati and Neb Jmel are an important quality parameter because they correlate with colour, which is the first attribute of virgin olive oil evaluated by consumers. In the studied oils, chlorophyllic and carotenoid pigments were found to be at concentrations between 1.13 ± 0.02 to 14.71 ± 0.55 mg/kg and 1.36 ± 0.02 to 5.83 ± 0.18 mg/kg, respectively (Table 1), with the highest amounts observed in the oils of southern Neb Jmel. The data have proven the influence of geographical location on carotenoid content. These results are in accordance with those reported by Issaoui et al., (2010), who claimed that there are significant differences between the Chemlali olive oils when grown in different environments [north (Béja: 222 m; 36°44'N, 09°11'E) and south (Sfax: 13 m; 34°44'N, 10°36'E) of Tunisia].

Pigment Content

The total pigment content of olive oils is an important quality parameter because it correlates with color, which is the first attribute of virgin olive oil evaluated by consumers.

Table 1 shows that the total amount of chlorophylls did not exceed 15 mg/kg for both Neb Jmel, Chemlali and Oueslati cultivars. Chlorophyll pigments have been found at concentration levels between 1.13 and 14.71 mg/kg (Table 1). The southern Neb Jmel oils contained the highest amounts, while the Oueslati had the lowest value. The growing area conditions also had a significant effect on the amount of chlorophylls: olive oils from the South had the highest level (11.30 and 14.71 mg/kg for Southern Chemlali and Neb Jmel olive oils, respectively). The same phenomenon was observed with carotenoids. Data proved the influence of geographical location on carotenoids content: both Chemlali and Neb Jmel olive oils from the South contained the highest levels 4.55 and 5.83 mg/kg, respectively. Carotenoids pigments varied between 1.89 and 5.83 mg/kg, with the lowest amounts observed in the oils of Oueslati (Issaoui et al., 2010; Ceci and Carelli, 2007) (Table 1).

Fatty Acid Composition

The fatty acid composition has a relatively wide range due to genetic and environmental factors. It has previously been used by a number of authors as a parameter for oil classification (Haddada et al., 2008). The mean FAME composition of the oils from the analysed cultivar is shown in Table 2. The major fatty acids were oleic, palmitic, linoleic, stearic and palmitoleic acids. As can be seen, the fatty acid composition of the studied oils is variable, depending on the place of origin (Table 2). The palmitic acid (16:0) mean value Chemlali cultivar was 19.46% in the south of Sfax, Oueslati 17.73% and of the Neb Jmel cultivar was 19.81% in the South. Oleic acid (18:1n-9), linoleic acid (18:2n-6), and stearic acid (18:0) were the main fatty acids in all olive oils. The other fatty acids were detected in low percentages. Monounsaturated isomers of fatty acids with sixteen carbon atoms (hypogeic acid: 16: 1n-9 and palmitoleic acid: 16: 1n-7) and eighteen carbon atoms (oleic acid: 18: 1n-9 and z-vaccenic acid: 18: 1n-7) are considered separately contrary to the commercial standards (trade standard) of the IOC (2013). The determination of these minor fatty acids is very important for the accurate knowledge of lipid profiles during the authentication of the varietal origin of olive oils (Olivier et al., 2003). The ratio of monounsaturated to polyunsaturated fatty acids (MUFA/PUFA) varied from 2.89 to 3.32% for Chemlali (Sfax) and Oueslati, respectively. the results described by (Aguilera et al., 2005) and also with those of Osman et al., (2004), indicating that the percentage of unsaturated fatty acids in olive oil increases with the decrease in temperature or increase in altitude. Apart from altitude and temperature, other climatic variables such as the soil characteristics of the olive grove zones and the salinity influence the chemical composition of VOO (Servili et al., 2004).

Table 1. Quality indices, Total phenols, flavonoid composition in Tunisian virgin olive oils samples

	Chemlali	Oueslati	Neb Jmel
Free acidity (%)	0.33 ± 0.00	0.31 ± 0.00	0.19 ± 0.0
K_{232}	2.20 ± 0.00	2.18 ± 0.10	1.66 ± 0.04
K_{270}	0.17 ± 0.01	0.18 ± 0.00	0.15 ± 0.00
PV (meq O_2/kg)	13.81 ± 3.53	13.25 ± 0.39	12.90 ± 0.37
Chlorophyll (ppm)	11.30 ± 1.56	1.13 ± 0.05	14.71 ± 0.55
Carotenoids (ppm)	4.55 ± 0.40	1.89 ± 0.05	5.83 ± 0.18

Each value reprlesents the mean of three determinations (n = 3) ± standard deviation K_{232} and K_{270} spectrophotometric indices, PV peroxide value.

Table 2. Fatty acid composition (%) evaluated in the Tunisian VOO samples in different geographical regions

	Chemlali	Oueslati	Neb Jmel
Palmitic acid (C16:0)	19.46 ± 0.53	17.73	19.81 ± 0.33
Hypogeic (16:1n-9) + palmitoleic (16:1n-7) acids	2.74 ± 0.05	2.28	3.13 ± 0.05
Heptadecanoic acid (C17:0)	0.05 ± 0.00	0.04	0.05 ± 0.00
Heptadecenoic acid (C17:1n-8)	0.07 ± 0.00	0.07	0.05 ± 0.00
Stearic acid (C18:0)	2.71 ± 0.05	2.30	3.93 ± 0.07
(Z)-Vaccenic (18:1n-7) + oleic (18:1n-9) acids	56.13 ± 0.96	56.28	59.87 ± 0.97
Linoleic acid (C18:2n-6, n-9)	17.46 ± 1.16	19.69	12.47 ± 0.21
Linolenic acid (C18:3n-3, n-6, n-9)	0.48 ± 0.01	0.66	0.73 ± 0.01
Arachidic acid (C20:0)	0.20 ± 0.00	0.41	0.53 ± 0.01
C18:1/C18:2	0.70 ± 0.03	2.86	4.69 ± 0.16
Gondoı̈c acid (C20:1n-9)	3.21 ± 0.82	0.18	0.15 ± 0.00
∑ SFA (%)	22.42 ± 0.58	20.66	24.52 ± 0.41
∑MUFA (%)	59.64 ± 1.04	58.81	61.79 ± 1.02
∑PUFA (%)	17.94 ± 1.17	20.35	74.99 ± 1.24
∑ MUFA/∑ PUFA	3.32 ± 0.88	2.89	-

Each value represents the mean of three determinations (n = 3) ± standard deviation SFA saturated fatty acids, MUFA monounsaturated fatty acids, PUFA polyunsaturated fatty acids.

Sterol Composition

The composition of the sterol fraction of olive oil is a very useful parameter for detecting adulterations or to check authenticity, since it can be considered as a fingerprint (Ben Temime et al., 2008). As shown in Table 3, the amounts of individual sterols vary according to the production area. In the chemlali, Oueslati and Chetoui variety, the highest phytosterol levels were found for *β*-sitosterol, followed by *D-5*-avenasterol, characteristic of the virgin olive oil in the pulp of the olive (Cornforth 2002). These two major sterols were strongly and negatively correlated, and there was a clear differentiation between sites (Table 3). However, small amounts of cholesterol, *D-7*-stigmasterol and *D-7*-avenasterol were also found in all samples. These results are in good agreement with data published elsewhere (Alves et al., 2005). Authors have reported that *β*-sitosterol, *D-5*-avenasterol and campesterol are the most representative sterols in virgin olive oils from the main Spanish, Portuguese and French cultivars. *β*-sitosterol is known to be the most abundant phytosterol in olive oil (Table 3).

Regarding *D-5*-avenasterol content, Chemlali, Chetoui and Oueslati olive oil were showed the value (5.79, 7.52 and 15.60%, respectively). The percentages of stigmasterol were lower than those of campesterol and the apparent *D-5*-avenasterol content. The analysed samples contained low levels of this sterol, which is indicative that the oil comes from healthy fruit (Ben Temime et al., 2008). The stigmasterol content is significantly affected by the geographical origin. All the olive oil samples analysed showed low campesterol content, with a total range from 2.66 to 3.57% (Chetoui and Oueslati). Campesterol content was below the threshold established by EU Regulations (4%). Practically no statistically significant differences were observed in the campesterol content in relation to the geographical region. These results are in agreement with those of other research studies (Ben Temime et al., 2008). As regards other authenticity indices established by the current legislation (European Union Commission Regulation (EEC) 1989/2003), cholesterol percentages in the studied samples were below the established limits of 0.5% and the percentages of stigmasterol were much lower than those of campesterol of the samples analysed (Table 3).

Erythrodiol and Uvaol Content

The triterpenic dialcohols (erythrodiol and uvaol) are also part of the unsaponifiable fraction of the olive oil. The sum of erythrodiol and uvaol in all studied samples was lower than the established limit of 4.5% for the "extra virgin" olive oil category. Among the studied samples, Chemlali virgin olive oils had the highest value of erythrodiol and uvaol (2.05%), whereas virgin olive oil extracted from Oueslati recorded the lowest value (1.76%) (Table 3). Total erythrodiol and uvaol have been proposed as applicable to the characterization of olive oils, since the sterol profile differs from one location to another and from one variety to another. These dialcohols are generally located in the exocarp of the olive (Ben Temime et al., 2008). Several researchers have described the use of olive oil's sterol profile to detect possible fraudulent admixtures. The presence of olive pomace in virgin oil can be detected from the levels of erythrodiol + uvaol (Reina et al., 1997). Total erythrodiol and uvaol have been proposed as applicable to the characterization of olive oils, since the sterol profile differs from one variety to another.

Table 3. Sterol and triterpenic dialcohols composition (%) of Tunisian virgin olive oils samples

	Chemlali	Oueslati	Chetoui
Cholesterol	0.25 ± 0.01	0.48 ± 0.02	0.24 ± 0.01
Campesterol	3.13 ± 0.27	3.57 ± 0.09	2.66 ± 0.23
Stigmasterol	0.66 ± 0.13	0.84 ± 0.03	3.36 ± 0.04
β-sitosterol	86.89 ± 0.35	70.57 ± 1.39	83.15 ± 0.70
D-5-avenasterol	5.79 ± 0.08	15.60 ± 0.25	7.52 ± 0.17
D-7-stigmasterol	0.43 ± 0.02	0.58 ± 0.16	0.19 ± 0.02
D-7-avenasterol	0.80 ± 0.10	0.55 ± 0.04	0.23 ± 0.02
Uvaol + erythrodiol	2.05 ± 0.06	1.76 ± 0.24	1.96 ± 0.15

Each value represents the mean of three determinations (n = 3) ± standard deviation.

Contents of Total Phenols and Flavonoids

The phenolic compounds present in virgin olive oils are one of the bases of the nutritional importance and shelf life of this oil (Haddada et al., 2008). It is important to note that northern cultivars Neb Jmel had the highest concentration of phenols 513 ± 18.9 (ppm), while the southern Chemlali oil recorded the lowest concentration 122.94 ± 18.9 (ppm). Lower mean values were found in Dhokar olives, they attained 1.45 ± 0.52 (ppm) of fresh olive fruit. These results are in accordance with some reports by different authors that the amount of phenolic compounds in olive oil varies from 50 to 1000 mg/kg depending on place of origin, variety, processing systems, etc. (Abaza et al., 2005; Baiano et al., 2013; Taamalli et al., 2010).

Total flavonoids content was also measured in the extracts. The mean values of total flavonoids content in the six cultivars studied are presented in Table 4. Chemlali showed significantly the highest mean content in total flavonoids (11.80 ± 1.45 ppm). However, flavonoid concentrations were lower in Chetoui, Zalmati, Chemchali, Dhokar, Neb Jmel and Gemri-dhokar cultivars and ranged from 0.23 to 2.09 ppm.

Table 4. Total phenolic content and total Flavonoid content in the seven varieties

	Total phenols (ppm)	Total flavonoids (ppm)
Chemlali	122.94 ± 0.70	11.80 ± 1.45
Chetoui	6.05 ± 0.85	0.23 ± 0.08
Zalmati	5.45 ± 0.70	0.23 ± 0.07
Chemchali	9.09 ± 1.15	0.26 ± 0.05
Dhokar	1.45 ± 0.52	0.52 ± 0.08
Gemri-dhokar	6.97 ± 0.67	2.09 ± 0.41
Neb Jmel	513.57 ± 18.9	2.06 ± 0.08

Each value represents the mean of three determinations (n = 3) ± standard deviation.

Quantitative Characterization of Olive Oil Phenolic Compounds

The analysis of phenolic substances using HPLC-DAD has allowed the separation and the identification of 15 phenolic compounds. The results have shown significant quantitative differences in the phenolic fraction between virgin olive oils from different geographical regions in a wide number of phenolic compounds. The most important secoiridoids detected were oleuropein aglycon (3,4-DHPEA-EA), Ligstroside aglycone dialdehyde (p-HPEA-EDA), oleuropein aglycone (3,4-DHPEA-EA)and decarboxymethylated derivates of oleuropein aglycon (3,4-DHPEA-EDA). *3,4*-DHPEA-EDA and *3,4*-DHPEA-EA are esters of hydroxytyrosol, while *p*-HPEA-EDA is an ester of tyrosol since the two sets of compounds are formed by hydrolysis of oleuropein and ligstroside, respectively (de Medina et al., 2015). The oxidation products of dialdehydic form of decarboxymethyl ligstroside aglycone (*p*-HPEA-EDA) were detected in low concentrations in Chemlali olive oil from Sfax with 1.12 mg kg^{-1}. Another, Fakharri cultivar showed the highest total content of phenolic compounds, being also the richest in terms of oleuropein aglycone (66.12 mg kg^{-1}). Indeed, secoiridoids concentration in the selected cultivars decreased in the following order: Fakharri > Jemri > Touffehi > Zalmati > Chemlali (Table 5). Concerning the amounts of phenyl alcohols, 3,4-DHPEA, *3,4*-DHPEA-AC and *p*-HPEA (hydroxytyrosol, hydroxytyrosol acetate and tyrosol, respectively) were the most abundant compounds quantified of this family. These three simple phenols were previously described in Tunisian EVOO especially in 'Chemlali,' 'Chétoui' and Oueslati cultivars (Ouni, 2014; Ouni et al., 2011). 3,4- DHEPA-Glu, a derivative compound of hydroxytyrosol (hydroxytyrosol glucoside), was also quantitated and found to reach 14.49 mg kg^{-1}. Therefore, significant differences between geographical areas were observed with regard to phenolic compounds. These results are similar to those reported by several authors for other olive oil varieties (Krichene et al., 2009). The reason for the wide content distribution observed for these two phenols may be that hydroxytyrosol and tyrosol can be produced by the partial hydrolysis of their derivatives (Montedoro et al., 1992). In addition, the high antioxidant activity of hydroxytyrosol, which is much higher than that observed for tyrosol (Papadopoulos et al., 1991), favours its degradation and disappearance. Lignans were also found in considerable amounts, particularly acetoxypinoresinol and pinoresinol. Concerning acetoxypinoresinol, it was found in analysed sample at concentrations 16.11 mg kg^{-1}. Regarding flavonoids such as luteolin and apigenins, they were quantitated in the olive oils studied. The concentration of luteolin was higher than that of apigenin. The major concentrations of the former were detected in Zalmati (15.59 mg kg^{-1}), whilts Chemlali oil exhibited the lowest levels (5.93 mg kg^{-1}) and the low concentration for apigenin was observed (0.09 mg kg^{-1}). These results confirm this chemical class as a minor constituent of the polyphenol fraction, as previously described for other virgin olive oil (Garcıa-Villalba et al., 2010). The olive oil obtained from the five demarcated regions has shown differences pertaining to the amounts of secoiridoids, phenolics alcohols, lignans and flavonoids. These differences could be attributed to the climatic and edaphological characteristics between the regions studied (Vekiari et al., 2010). The effects of these variables on the phenolic composition have also been mentioned by other authors (Tura et al., 2007; Ocakoglu et al., 2009; Alkan et al., 2012; Ouni et al., 2011; Baiano et al., 2013; Taamalli et al., 2010).

Antioxidant Activity DPPH Assay

IC_{50} values of the five studied cultivars are shown in Table 6. Chemlali oils have the lowest value (16.24 μg/mL) followed by Fakharri, Touffehi, Zalmati, Jemri and Neb Jmel oils. This ranking can be explained through antioxidant activity correlation to polyphenols concentration and chemical structure, and consequently to their free radical scavenging capacity and metal chelating properties (Bouaziz et al., 2004). Therefore, the number of hydroxyl groups and their position are important for the antioxidant capacity of phenolic compounds, which is linked to the formation of stable phenoxyl radicals. As the matter of fact, especially hydrophilic phenols, such as phenolic acids, phenolic alcohols, flavonoids and secoiridoids, are responsible of the high antioxidant activity. In addition, oleuropein aglycone and hydroxytyrosol are the most important antioxidants in olive oil (Bulotta et al., 2014). The reduction of the concentration in these two compounds during the ripening period could lead to a decrease in the antioxidant activity. Recently, antioxidants have become a topic of increasing interest to health and food science researchers or medical experts (Rigane et al., 2011).

Table 5. Phenolic compound compositions (mg/kg ± standard deviation) of five Tunisian olive oil samples

Phenolic compound	Chemlali	Jemri	Touffehi	Fakharri	Zalmati
3,4-DHPEA	19.79 ± 0.16	1.78 ± 0.01	1.68 ± 0.02	1.24 ± 0.01	3.09 ± 0.03
3,4-DHEPA-Glu	14.49 ± 0.14	-	-	-	-
3,4-DHEPA-AC	-	0.24 ± 0.00	0.15 ± 0.00	0.15 ± 0.00	0.28 ± 0.00
p-HPEA	16.04 ± 0.11	0.25 ± 0.00	0.34 ± 0.00	0.34 ± 0.00	10.99 ± 0.04
Cafeic acid	0.78 ± 0.38	0.35 ± 0.00	0.97 ± 0.01	0.53 ± 0.00	0.21 ± 0.00
Coumaric acid	1.49 ± 0.21	3.28 ± 0.02	1.77 ± 0.02	1.67 ± 0.02	1.63 ± 0.02
Ferulic acid	1.52 ± 0.17	0.01 ± 0.00	0.01 ± 0.00	0.01 ± 0.00	0.06 ± 0.00
p-HPEA-EA	4.09 ± 0.36	-	-	-	-
3,4-DHPEA-EA	43.69 ± 0.68	60.22 ± 0.51	53.37 ± 0.47	66.12 ± 0.49	52.66 ± 0.44
p-HPEA-EDA	1.12 ± 0.52	55.31 ± 0.33	46.11 ± 0.33	64.22 ± 0.40	47.85 ± 0.40
3,4-DHPEA -EDA	3.32 ± 0.41	42.02 ± 0.38	45.72 ± 0.32	58.20 ± 0.38	48.67 ± 0.41
Pinoresinol	23.21 ± 0.28	-	-	-	-
Acetoxy-pinoresinol	16.11 ± 0.85	-	-	-	-
Luteolin	5.93 ± 0.22	15.54 ± 0.13	16.27 ± 0.19	13.13 ± 0.15	15.59 ± 0.1
Apigenin	0.09 ± 0.01	1.72 ± 0.01	2.95 ± 0.03	1.63 ± 0.01	5.72 ± 0.04

Each value represents the mean of three determinations (n = 3) ± standard deviation.

*Identification confirmed by comparison with standards.3,4-DHPEA Hydroxytyrosol, 3,4-DHPEA-Glu Hydroxytyrosol Glucoside, 3,4-DHPEA-AC Hydroxytyrosol Acetate, p-HPEA Tyrosol, p-HPEA-EA Ligstroside aglycon, 3,4-DHPEA-EA Oleuropein aglycon, p-HPEA-EDA Dialdehydic form of elenolic acid linked to Tyrosol, 3,4-DHPEA-EDA Dialdehydic form of elenolic acid linked to Hydroxytyrosol.

Sensorial Evaluation

The results of the sensory analysis of the oil samples are shown in Figure 1. Three positive descriptors (fruitiness, bitterness and pungency) and the defects (mustry and rancid). The results indicated that the intensities of these attributes varied according to the cultivars and the maturity index. The sensory analysis of Chemlali and Oueslati oils obtained from the whole fruits and showed a moderately intense fruity flavor. Indeed, the sensory profiles of Neb Jmel olive oils were fruity, with attributes, such as pungent and bitter, varying only slightly in intensities.

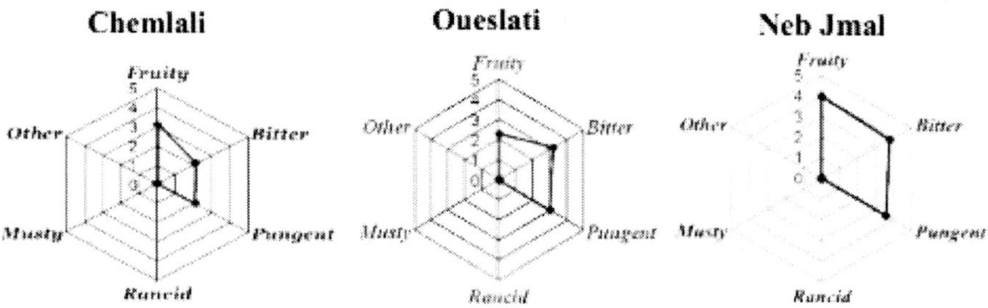

Figure 1. Sensory wheels of virgin olive oils for the Tunisian olive cultivars chemlali growing in Sfax regions, Oueslati and Neb Jmel.

Table 6. Oxidative stability and IC$_{50}$ of the different olive oil samples during sampling times

	Chemlali	Neb Jmel	Jemri	Touffehi	Fakhari	Zalmati
Oxidative stability (h)	5.61 ± 0.38	8.34 ± 0.15	9.23 ± 0.54	8.25 ± 0.15	14.26 ± 0.92	13.31 ± 1.10
IC50 (µg/ml)	16.24 ± 0.00	24.14 ± 10.9	150.2 ± 11.9	130.5 ± 11.6	120.1 ± 10.9	145.1 ± 10.1

Each value represents the mean of three determinations (n = 3) and three independent experiments ± standard deviation. IC50 median inhibitory concentration (concentration at which an antagonist exerts its half-maximal effect).

Well-balanced bitter and pungent tastes are due in particular to the presence of oleuropein aglycon (Mateos et al., 2004), and deacetoxyligustroside aglycon (Andrewes et al., 2003), with defect median equal to zero. According to the sensory evaluation, the oils under investigation belong to the extra virgin olive oil category. In fact, a reduction in positive attributes, especially fruity flavor were apparent. It became clear that sensory quality was the result of the synergistic effect of the various oil components, which are influenced by the pedoclimatic conditions, ripening index and part of olive fruits used to extract the oil (Issaoui et al., 2010; Gargouri et al., 2013).

Table 7. Tocopherols (mg/kg) of the different olive oil samples

	Jemri	Touffehi	Fakharri	Zalmati
α-Tocopherols (mg/kg)	263.0 ± 12.1	278.1 ± 14.0	303.0 ± 12.1	350.1 ± 18.1
β-Tocopherols (mg/kg)	4.3 ± 0.2	4.9 ± 0.1	5.0 ± 0.1	7.0 ± 0.2
γ-Tocopherols (mg/kg)	6.5 ± 0.2	9.7 ± 0.1	7.8 ± 0.1	9.0 ± 0.2

Each value represents the mean of three determinations (n = 3) and three independent experiments ± standard deviation.

Oil Stability

Stability to oxidation is an important property of olive oil, which is improved by synergistic interactions between the various antioxidants present in the oil itself, and depends on the lipid composition. Oxidative stability varied according to the growing area and cultivar. It ranged from a minimum of 5.61 h for southern Chemlali to a maximum of 14.26 h for Fhakkarri (Table 6). The high stability of virgin olive oil against thermo-oxidation and autoxidation processes over time is correlated to the antioxidant effects of polyphenolic compound such as oleuropein aglycone and hydroxytyrosol, (Tura et al., 2007). Indeed, the lowest stability value of Chemlali olive oils may explained by the lower level in phenols (<200 ppm) which showed also a lower antioxidant activity. However, the highest stability was registered in Fakharri oils, which can be explained by its richness in phenol and α-tocopherols contents (208.8 and 303.0 mg kg^{-1}, respectively).

Tocopherols Composition

The tocopherol concentration in olive oils mainly depends on the cultivar and the degree of fruit maturation (Ballus et al., 2014; Jimenez et al., 2015). Three tocopherol isomers were detected and quantified in the Tunisian EVOO samples which were: α-tocopherol, β-tocopherol and γ-tocopherol. Table 7 presents the results of all EVOO samples as well as a comparison between tocopherol contents of the four studied varieties during repining. The α-tocopherol isomer was quantified in all samples and ranged between 263 and 350 mg kg^{-1}, while β-tocopherol ranged from 4.3 to 7.0 mg kg^{-1} and γ-tocopherol ranged between 6.5 and 9.7 mg kg^{-1}. Since α-tocopherol is the main tocopherol isomer found in EVOOs, the most promising Tunisian varieties with respect to its content were Zalmati and Fakharri which exhibited high contents surpassing 300 mg kg^{-1}.

Volatile Compound Analyses

The fragrance and unique flavor of EVOO represents some of the most important qualitative aspects of this vegetable oil, and plays a major role in the consumer approval.

Although a full description of the organoleptic characteristics of the oil is only obtainable through sensory analysis, the qualitative and quantitative determination of the volatile compounds can provide very useful information on the product quality (Ammar et al., 2014). In order to evaluate the matrix volatile composition, HS-SPME was used. Twenty-four volatile compounds (Table 8) were identified by the GC-MS analysis. Table 8 shows the influence of cultivar and growing area on the percentage of volatile compounds obtained from the oils of different samples harvested at the same degree of ripeness and processed under the same operating conditions. These compounds represent 95.06 - 99.52% of the total amount. The number of aromatic compounds differed according to the environmental conditions: olive oils from the cultivars produced a higher number of aromatic compounds for the southern Neb Jmel, Chemlali,

Oueslati and Chetoui (Table 8). The number also differed according to the cultivar: olive oils from the Neb Jmel cultivar produced a higher number of aromatic compounds, 15 in comparison to the Chemlali, Oueslati and Chetoui oils. These results indicate that olive oil aroma compounds accumulate differently according to the geographic area and the cultivar kind (Issaoui et al., 2010; Baccouri et al., 2008). These results are in agreement with findings of Nigri et al., 2012 in Algerian olive oils. For the oil samples involved in this study, (*E*)-2-hexenal was the major volatile in approximately 60% of the tested oils (Table 8). The major constituents of the volatile fraction of Chemlali South oils were (*E*)-2-hexenal (54.61%), (*E*)-2-dodecene (10.11%), 1-hexanol (6.87%), *n*-dodecane (6.09%), 3,7-decadiene(4.51%)and *n*-tetradecane (4.02%)(Table 8). It is important to note that (*E*)-2-dodecene, present in most all the studied oils, was significantly more abundant in Chemlali olive oil (10.11%) than in Neb Jmel olive oil (3.59%) and Oueslati (1.0%). The major constituents of the volatile fraction of the southern Neb Jmel oils were (*E*)-2-hexenal (64.57%), 1-hexanol (10.13%), 3,7-decadiene (6.04%) and (*E*)-2-dodecene (3.59%). The percentage of esters compounds, especially (*Z*)-3-hexenyl acetate and 1-hexyl acetate, which were absent in Chemlali and Neb Jmel from the south and present in Oueslati and Chetoui. Taking into consideration the effect of the growing area conditions, it is important to notice that the cultivars grown at lower altitudes and higher temperature (south) had the highest level of alcools; 1-hexanol was not detected in the cultivars grown at higher altitude and lower temperature (center). Hence, we can hypothesize that 1-hexanol (1.2%, 6.87%, 10.13% and 19.9% for Oueslati, Chemlali South, Neb Jmel South and Chetoui, respectively) may be used as a marker to differentiate VOO of different geographical origins. The second aroma represented by Ouni et al., (2010), 19 compounds-richest was that of Oueslati variety, in which (E)-2-hexanal was the dominant one (20.9%) followed by hexanal (10.2%), (E,E)- α-farnesene (6.8%) and (Z)-3-hexenyl acetate (5.9%). In the Chetoui olive oil headspace, 18 different compounds were identified, accounting for 96.8% of total volatiles.

Table 8. Composition of volatiles obtained from different olive oil samples according to their geographical origin

	Volatil Compounds	L.r.iA	L.r.iB	Chemlali	Neb Jmel	Oueslati	Chetoui
1	Hexanal	802	1071	-	-	10.2 ± 0.1	3.7 ± 0.0
2	(E)-2-hexenal	856	1222	54.61 ± 1.45	64.57 ± 2.29	20.9 ± 0.1	24.1 ± 0.3
3	1-hexanol	869	1352	6.87 ± 0.41	10.13 ± 0.83	1.2 ± 0.0	19.9 ± 0.2
4	(Z)-4-heptenal	902	1250	-	0.82 ± 0.09		
5	3-ethyl-1,5-octadiene	942	1008	-	1.63 ± 0.07	-	-
6	3-ethyl-1,5-octadiene	950	1016	-	1.78 ± 0.15	-	-
7	Benzaldehyde	964	1494	0.62 ± 0.08	-	-	-
8	β-pinene	982	1108	-	-	-	3.04 ± 0.0
9	3,7-decadiene (isomer not identified)	999	1074	4.51 ± 0.18	6.04 ± 0.37	-	-
10	(Z)-3-hexenyl acetate	1008	1310	-	-	5.9 ± 0.0	13.0 ± 0.1
11	1-hexyl acetate	1010	1269	-	-	3.8 ± 0.1	3.7 ± 0.0
12	1,8-cineole	1034	1209	1.31 ± 0.05	-	-	-
13	butyl 2-methylbutanoate	1045	1230	0.89 ± 0.02	-	-	-
14	(E)-β-ocimene	1052	1255	-	0.19 ± 0.02	1.2 ± 0.1	2.4 ± 0.1
15	(E)-2-octenal	1062	1343	-	0.79 ± 0.08	-	-
16	nonanal	1104	1382	1.72 ± 0.04	0.62 ± 0.05	2.9 ± 0.4	-
17	2-methyl-6-methylene-1,7-octadien-3-one	1110	-	2.48 ± 0.07	-	-	-
18	Camphor	1145	1522	1.01 ± 0.03	2.91 ± 0.05	0.4 ± 0.09	0.37 ± 0.0
19	n-dodecane	1200	1200	6.09 ± 0.35	4.12 ± 0.24	2.7 ± 0.3	-
20	(E)-2-dodecene	1206	1263	10.11 ± 0.04	3.59 ± 0.03	1.0 ± 0.3	-
21	(E)-2-decenal	1263	1590	-	-	-	-
22	n-tridecane	1300	1300	0.82 ± 0.01	0.32 ± 0.04	0.4 ± 0.2	-
23	n-tetradecane	1400	1400	4.02 ± 0.19	1.69 ± 0.08	1.5 ± 0.2	1.8 ± 0.1
24	(E,E)-α-farnesene	1508	1727	-	0.32 ± 0.01	4.0 ± 0.4	7.4 ± 0.2

Table 8. (Continued)

Volatil Compounds	L.r.iA	L.r.iB	Chemlali	Neb Jmel	Oueslati	Chetoui
Monoterpene hydrocarbons			1.61	0.49	0.5	-
Oxygenated monoterpenes			1.19	1.09	0.4	-
Sesquiterpene hydrocarbons			0.00	0.61	19.8	-
Non-terpene derivatives			94.23	97.00	75.8	-
Total identified			95.06	99.52	96.5	96.8
Totals Alcohols			0.00	10.12	-	-
Totals Aldehydes			62.13	65.44	-	-
C6 Alcohols			0.00	10.13	-	-
C6 Aldehydes			62.12	64.63	-	-

A: apolar column; B: polar column; L.r.i: linear retention index; each value represents the mean of three determinations (n = 3) ± standard deviation.

It contained (E)-3-hexanol, 1-hexanol, decanal, (E)-2- hexenal, (Z)-3-hexenyl acetate and (E,E)-α-farnesene among its major components (26.1, 19.9, 14.2, 13.0, 7.4 and 19.0%, respectively) (Table 8). It is also important to note the increasing level of 1-hexanol (detected at considerable lower levels in Oueslati), which emphasizes the perception of green pleasant attributes. The volatile compounds are not only affected by the cultivars and growing area but also they are extraction-system dependent (Volpe et al. 2014). All the oils obtained in the laboratory contain volatile compounds derived from the biochemical pathway of lipoxygenase, which allows the catalytic oxidation of 1–4 pentadienic structures to give C6 derivatives (Flamini et al., 2003; Williams et al., 2000; Ridolfi et al., 2002). In particular, (*E*)-2-hexenal and hexanal are derived from the heterolytic breakdown of the 13-hydroperoxide of linoleic and linolenic acid, respectively, catalyzed by enzymes involved in the reaction of hydroperoxide lyase in accordance with published data (Olias et al., 1993). In oil obtained from the southern Chemlali and Neb Jmel cultivars, the high level of 1-hexanol can be attributed to the increased presence or activity of the Alcohol Dehydrogenase (ADH) enzyme. ADH has previously been reported as the enzyme responsible for the biosynthesis of six-carbon alcohols, such as 1-hexanol (Olias et al., 1993).

CONCLUSION

This is the first time in which a complete characterization of oils, phenolic and tocopherols composition of rare Tunisian cultivars such as Jemri, Touffehi, Fakharri and Zalmati are described. The use of LC-ESI-MS/MS in MRM mode allowed a correct identification and a reproducible quantification for eleven phenolic compounds, which one of them was identified for the first time in Tunisian olive oils (Diosmetin). Phenolic compounds displayed significant qualitative and quantitative differences among the cultivars and their sampling times. However, we observed that Zalmati oil stood out for its flavonoids and simple phenols content. Whereas Jemri and Touffehi showed high levels of phenolic acids, while Fakharri and Chemlali oils for its secoiridoids compounds. A clear difference between cultivars was observed for the oleuropein aglycone content which is the main secoiridoid component found in the phenolic profile. These two minor components affected directly the antioxidant properties of the studied oils. Typical sensory properties of EVOO, such as bitterness and pungency, have been attributed to secoiridoid derivatives of oleuropein and demethyloleuropein such as *3,4-* DHPEA-EDA and *3,4-* DHPEA-EA. In addition, the results of sterol fractions shows that β-sitosterol, *D-5*-avenasterol, stigmasterol and the sum of erythrodiol and uvaol are variables for characterizing the chemlali, Oueslati and Chetoui olive oils. However, the application of SPME to the analysis of virgin olive oil headspace allowed the detection of significant differences in the proportion of volatile constituents from oils of different cultivars. Chemlali olive oil showed a low tolerance against the variation in climate and plantation zone. Neb Jmel olive oil revealed excellent nutritional characteristics, in terms of antioxidant compounds. The results indicate that the cultivar and the environmental conditions influence the quali-quantitative production of volatiles.

REFERENCES

Abaza, L., Taamalli, W., Ben Temime, S., Daoud, D., Gutierrez, F., Zarrouk, M. (2005). Natural antioxidant composition as correlated to stability of some Tunisian virgin olive oils. *J Riv Ital Sostanze Grasse.* 82, 12–18.

Adams, Robert P. (1995). *Identification of Essential Oil Components by Gas Chromatography/Mass Spectrometry.* Carol Stream, IL: Allured Publishing Corp.

Adams, R. P., (2003). Identification of essential oil components by gas chromatography/mass Spectroscopy. *Allured: Carol Stream, IL.*

Aguilera, M. P., Beltra´n, G., Ortega, D., Ferna´ndez, A., Jime´nez, A., Uceda, M. (2005) Characterisation of virgin olive oil of Italian olive cultivars: 'Frantoio' and 'Leccino,' grown in Andalusia. *Food Chem.* 89, 387–391.

Alkan, D., Tokatli, F., Ozen, B. (2012) Phenolic characterization and geographical classification of commercial extra virgin olive oils produced in Turkey. *J Am Oil Chem Soc* 89, 261–268.

Alves, MR., Cunha, SC., Amaral, JS., Pereira, JA., Oliveira, MB. (2005). Classification of PDO olive oils on the basis of their sterol composition by multivariate analysis. *Anal Chem Acta* 549, 166–178.

Ammar, S., Zribi, A., Gargouri, B., Flamini, G., Bouaziz, M. (2014). Effect of addition of olive leaves before fruits extraction process to some monovarietal Tunisian extra virgin olive oils using chemometric analysis. *J. Agric. Food Chem.* 50, 835-839.

Andrewes, P., Bush, J. L. H. C., De Joode, T., Groeneken, A., Alexandre, H.(2003) Sensory properties of virgin olive oil polyphenols: Identification of deacetoxy-ligstroside aglycon as a key contributor to pungency. *J. Agric. Food Chem.* 51, 1415–1420.

Angerosa, F. (2002). Influence of volatile compounds on virgin olive oil quality evaluated by analytical approaches and sensor panels. *Eur. J. Lipid Sci. Technol.* 104, 639–660.

Angerosa, F., Mostallino, R., Basti, C., Raffaella, V. (2000) Virgin olive oil odour notes: their relationships with volatile compounds from the lipoxygenase pathway and secoiridoid compounds. *Food Chem* 68, 283-287.

Aparicio, R., Luna, G. (2002) Characterisation of monovarietal virgin olive oils. *Eur. J. Lipid Sci. Technol* 104, 614–627.

Baccouri, B., Zarrouk, W., Khichene, D., Nouari, I., Ben Youssef, N., Daoud, D. (2007) Influence of fruit ripening and crop yield on chemical properties of virgin olive oils from Seven selected oleasters (Olea europea L.). *J Agron.* 6, 388–396.

Baccouri, O., Bendini, A., Cerretani, L., Guerfel, M., Baccouri, B., Lercker, G., Zarrouk, M., Ben-Miled, D (2008). Comparative study on volatile compounds from Tunisian and Sicilian monovarietal virgin olive oils. *Food Chem.* 111, 277-296.

Baiano, A., Terracone, C., Viggiani, I., Del Nobile, MA. (2013). Effects of cultivars and location on quality, phenolic content and antioxidant activity of extra-virgin olive oils. *J Am Oil Chem Soc.* 90, 103–111.

Ballus, C. A., Meinhart, A. D., de Souza Campos, F. A., da Silva, L. F. de O., de Oliveira, A. F., & Godoy, H. T. (2014). A quantitative study on the phenolic compound, tocopherol and fatty acid contents of monovarietal virgin olive oils produced in the southeast region of Brazil. *Food Research International*, *62*, 74–83.

Ben Brahim, S., Kelebek, H., Ammar, S., Abichou, M., Bouaziz, M (2016) LC-MS phenolic profiling combined with multivariate analysis as an approach for the characterization of extra virgin olive oils of four rare Tunisian cultivars during ripening. *J. Food Chem.* DOI. http://dx.doi.org/10.1016/j.foodchem.2017.02.025.

Ben Mansour, A., Flamini, G., Ben Selma, Z., Le Dréau, Y, Artaud J, Abdelhedi R, Bouaziz M (2015a) Olive oil quality is strongly affected by cultivar, maturity index and fruit part: Chemometrical analysis of volatiles, fatty acids, squalene and quality parameters from whole fruit, pulp and seed oils of two Tunisian olive cultivars. *Eur. J. Lipid Sci. Technol* 117, 976–987.

Ben Mansour A, Gargouri B, Flamini G, Bouaziz M (2015b) Effect of agricultural sites on differentiation between Chemlali and Neb Jmel olive oils. *J Oleo Sci* 64: 381-392.

Ben Mansour, A., Chtourou, F., Khbou, W., Flamini, G., Bouaziz, M. (2016). Phenolic and volatile compounds of Neb Jmel olive oil cultivar according to their geographical origin using chemometrics. *Eur Food Res Tech*. DOI 10.1007/s00217-016-2754-5.

Ben Temime, S., Manai, H., Methenni, K., Baccouri, B., Abaza, L., Daoud, D., Casas JS, Bueno, EO., Zarrouk, M. (2008). Sterolic composition of Che′toui virgin olive oil: influence of geographical origin. *Food Chem* 110, 368–374.

Bouaziz, M., Chamkha, M., & Sayadi, S (2004) Comparative study on phenolic content and antioxidant activity during maturation of the olive cultivar Chemlali from Tunisia. *Journal of Agricultural and Food Chemistry*, *52*(17), 5476–5481.

Bouaziz, M., Feki, I., Ayadi, M., Jemai, H., Sayadi, S. (2010) Stability of refined olive oil and olive pomace oil added by phenolic compounds from olive leaves. *Eur J Lipid Sci Technol* 112(8), 21–24.

Boskou Dimitrios (2006) *Olive Oil 2nd Edition. Chemistry and Technology*. Academic Press and AOCS Press.

Bulotta, S., Celano, M., Massimo Lepore, S., Montalcini, T., Pujia, A., and Diego Russo (2014). Beneficial effects of the olive oil phenolic components oleuropein and hydroxytyrosol: focus on protection against cardiovascular and metabolic diseases. *J Transl Med*. 12, 219.

Campeol, E., Flamini, G., Chericoni, S., Catalano, S., Cremonini R. (2001) Volatile compounds from three cultivars of Olea europea from Italy. *J. Agric. Food Chem.* 49, 5409-5411.

Ceci, L. N., Carelli, A. A. (2007). Characterization of Monovarietal Argentinian Olive Oils from New Productive Zones. *J. Am. Oil Chem. Soc.* 84, 1125-1136.

Chtourou, M., Gargouri, B., Jaber, H., Abdelhedi, R., & Bouaziz, M (2013) Comparative study of olive oil quality from Chemlali Sfax versus

Arbequina cultivated in Tunisia. *European Journal of Lipid Science and Technology*, *115*(6), 631–640.

Conseil oléicole international. 2012. Politique – Tunisie.

Cornforth, JW. (2002) Sterol biosynthesis: the early days. *Biochem Biophys Res Commun* 292, 1129–1138.

Criado, M N., Motilva, MJ., Goni Mand Romero, MP. (2005) Comparative study of the effect of maturation process of the olive fruit on the chlorophyll and carotenoid fractions of drupes and virgin oils from Arbequina and Farga cultivars. *Food Chem* 100, 748–755.

European Union Commission Regulation EEC 2568/91 (1991). On the characteristics of olive and olive pomace oils and their analytical methods. *Official J Eur Com* L248.

European Union Commission Regulation (EEC) 1989/2003 Characteristics of olive and olive pomace oils and their analytical methods. *Official J Eur Com* 295, 57–66.

European Community, Commission Regulation 702/2007 of 21 June (2007) amending Regulation (EC) No. 2568/91 on the characteristics of olive oil and olive pomace oil and on the relevant methods of analysis. *Official J Eur Com L*. 161, 11–17.

Flamini, G., Cioni, P. L., Morelli, I. (1995) Volatiles from leaves, fruits, and virgin oil from Olea europaea Cv. Olivastra Seggianese from Italy. *J. Agric. Food Chem*. 51, 1382-1386.

Flamini, R. (2003). Mass spectrometry in grape and wine chemistry. Part I: Polyphenols. *Mass spectrometry reviews*. 22, 4, 218–250.

Garcı´a-Villalba, R., Carrasco-Pancorbo, A., Oliveras-Ferraros, C., Va´zquez-Martı´n, A., Mene´ndez, JA., Segura-Carretero, A., Ferna´ndezGutie´rrez, A. (2010). Characterization and quantification of phenolic compounds of extra-virgin olive oils with anticancer properties by a rapid and resolutive LC-ESI-TOF MS method. *J Pharm Biomed Anal* 51(2), 416–429.

Gargouri, B., ammar, S., Zribi, A., Ben Mansour, A., Bouaziz, M. (2013) Effect of growing region on quality characteristics and phenolic compounds of Chemlali extra-virgin olive oils. *Acta Physiol Plant*. 35, 2801–2812.

Gharbi, I., Issaoui, M., Mehri, S., Cheraief, I., Sifi, S., & Hammami, M (2015) Agronomic and technological factors affecting Tunisian olive oil quality. *Agricultural Sciences*, 6, 513-526.

Gómez-Rico, A., Fregapane, G., Salvador, M D. (2008). Effect of cultivar and ripening on minor components in Spanish olive fruits and their corresponding virgin olive oils. *Food Res Inter* 41: 433–440.

Haddada, FM., Kriche`ne, D., Manai, H., Oueslati, I., Daoud, D., Zarrouk, M. (2008) Analytical evaluation of six monovarietal virgin olive oils from Northern Tunisia. *Eur J Lipid Sci Technol.* 110, 905–913.

International Olive Council, COI/T, 15/NC No 3/Rev 5 (2010) *Principe de Vergara*, 154, 28002, Madrid, Spain.

International olive council (2013). *Determination of the composition and content of sterols and triterpene dialcohols by capillary column gas chromatography*. coi/ t.20/ doc. no 30/rev. 1.

Issaoui, M., Flamini, G., Brahmi, F., Dabbou, S., Ben Hassine, K., Taamali, A., Chehab, H., Ellouz M, Zarrouk, M., Hammami, M. (2010). Effect of the growing area conditions on differentiation between chemlali and che´toui olive oils. *Food Chem.* 119. 220–225.

Jabeur, H., Zribi, A., Bouaziz, M. (2015) Extra-virgin olive oil and cheap vegetable oils: distinction and detection of adulteration as determined by GC and chemometrics. *Food anal method.*, DOI: 10.1007/s12161-015-0249-9.

Jimenez, B., Sánchez-Ortiz, A., Lorenzo, M. L., & Rivas, A. (2015) Effect of Agronomical Practices on the Nutritional Quality of Virgin Olive Oil at Different Ripening Stages. *Journal of the American Oil Chemists' Society*, *92*(10), 1491–1501.

Kiritsakis, AK. (1990) *Olive Oil.* AOCS Press, Champaign.

Krichene, D., Allalout, A., Baccouri, B., Fregapane, GQ., Salvador, MD., Zarrouk, M. (2009) Territorial investigation based on the chemical composition of chemlali virgin olive oils. *Asian J Biochem* 4, 1–12.

Luchetti, F. (2002) Importance and future of olive oil in the world market-an introduction to olive oil. *European Journal of Lipid Science and Technology*, 104, 559e563.

Mateos, R., Cert, A., Pérez Camino, M. C., Garcia, M. J. (2004) Evaluation of virgin olive oil bitterness by quantification of secoiridoid derivatives. *J. Am. Oil Chem. Soc.* 81, 71–75.

Medina, V. S., Priego-Capote, F., & de Castro, M. D. L (2015) Characterization of monovarietal virgin olive oils by phenols profiling. *Talanta*, *132*, 424–432.

Montedoro, GF., Servilli, M., Baldioli, M., Miniati, E. (1992). Simple and hydrolyzable phenolic compounds in olive oil. Their extraction, separation, and cuantitative and semicuantitative separation and evaluation by HPLC. *J Agric Food Chem* 40, 1571–1576.

Nigri, S., Oumeddour, R., Fernandez, X. (2012). Analysis of some Algerian virgin olive oils by headspace solid phase micro-extraction coupled to gas chromatography/ mass spectrometry. *Riv. Ital. Sostanze Grasse* 89, 54-61.

Ocakoglu, D., Tokatli, F., Ozen, B., Korel, F. (2009). Distribution of simple phenols, phenolic acids and flavonoids in Turkish monovarietal extra virgin olive oils for two harvest years. *Food Chem* 113,401–410.

Office National de l'huile. (2015) http://www.onh.com.tn/index.php/fr/commercialisation/positionnement-sur-le-marche-mondial.

Ollivier, D., Artaud, J., Pinatel, C., Durbec, J. P., Guérère, M (2003) Triacylglycerol and Fatty acid Compositions of French Virgin Olive Oils Characterization by Chemometrics. *J. Agric. Food Chem.* 51, 5723–5731.

Olias, J. M., Pérez, A. G., Rios, J. J., Sanz, L. C. (1993). Aroma of virgin olive oil: biogenesis of the 'green' odor notes. *J. Agric. Food Chem.* 41, 2368-2373.

Osman, M., Metzidakis, I., Gerazoupoulos, D., Kiritsakis, A. (2004). Qualitative changes in olive oil of fruits collected from trees grown at two altitudes. *Riv Ital Sost Grasse.* 71, 187–190.

Ouni, Y., Taamalli, A., Go´mez-Caravaca, AM., Segura-Carretero, A., Fernández-Gutiérrez, A, Zarrouk, M. (2011). Characterisation and quantification of phenolic compounds of extra-virgin olive oils according to their geographical origin by a rapid and resolutive LC–ESI-TOF MS method. *Food Chem* 127, 1263–1267.

Ouni, Y., Flamini, G., Issaoui, M., Ben Youssef, N., Pier Luigi, C., Hammami, M., Daoud, D., Zarrouk, M (2010). Volatile compounds and compositional quality of virgin olive oil from Oueslati variety: Influence of geographical origin. *Food Chem* 124, 1770-1776.

Papadopoulos, GK., Boskou, D. (1991) Antioxidant effect of natural phenols on olive oil. *J Am Oil Chem Soc* 68, 669–671.

Reina, RJ., White, KD., Jahngen, EGE. (1997). Validated method for quantitation and identification of 4,4-desmethyl-sterols and triterpene diols in plant oils by thin-layer-chromatography high resolution gas chromatography-mass spectrometry. *J AOAC Int* 80(6), 1272–1280.

Ridolfi, M., Terenziani, S., Patumi, M., Fontanazza, G. (2002) Characterization of the lipoxygenases in some olive cultivars and determination of their role in volatile compounds formation. *J. Agric. Food Chem.* 50, 835- 839.

Rigane, G., Ben Salem, R., Sayadi, S., Bouaziz, M. (2011). Phenolic composition, isolation and structure of a new deoxyloganic acid derivative from Dhokar and Gemri-Dhokar olive cultivars. *J Food Sci* 76(7), 965–973.

Servili, M., Selvaggini, R., Esposto, S., Taticchi, A., Montedoro, GF., Morozzi, G. (2004) Health and sensory properties of virgin olive oil hydrophilic phenols: agronomic and technological aspects of production that affect their occurrence in the oil. *J Chromatogr A.* 1054, 113–127.

Taamalli, A., Gomez-Caravaca, AM., Zarrouk, M., Segura-Carretero, A., Ferna´ndez-Gutie´rrez A. (2010). Determination of a polar and minor polar compounds and other chemical parameters for the discrimination of six different varieties of Tunisian extra-virgin olive oil cultivated in their traditional growing area. *Eur Food Res Technol* 231, 965–975.

Trigui, A., Msallem, M. (2002). Catalogue des variétés autochtones et types locaux. Ministère de l'Agriculture, *IRESA*, Institut de l'Olivier 159.

Tura, D., Gigliotti, C., Pedo, S., Failla, O., Bassi D., Serraiocco, A. (2007) Influence of cultivar and site of cultivation on levels of lipophilic and hydrophilic antioxidants in virgin olive oils (Olea europaea L.) and correlations with oxidative stability. *Scientia Horticul.* 112, 108–119.

Vekiari, S. A., Oreopoulou, V., Kourkoutas. Kamoun, M. Msallem., Psimouli and Arapoglou, D. (2010). Characterization and seasonal variation of the quality of virgin olive oil of the Throumbolia and Koroneiki varieties from Southern Greece. *Grasas Y Aceites*, 61 (3), 221-231.

Volpe, M. G., De Cunzo, F., Siano, F., Paolucci, M., Barbarisi, C.,; Cammarota, G. (2014) Influence of extraction techniques on physical-chemical characteristics and volatile compounds of extra virgin olive oil. *J. Oleo Sci.* 63, 875-883.

Williams, M., Salas, J. J., Sanchez, J., Harwood, J. L. (2000) Lipoxygenase pathway in olive callus cultures (Olea europaea). *Phytochemistry* 53, 13-19.

In: Olive Oil
Editors: T. Fritjof and B. Henning
ISBN: 978-1-53612-563-4

Chapter 4

THE ENRICHMENT OF MASHED POTATOES WITH EXTRA VIRGIN OLIVE OIL: THE EFFECTS ON TEXTURAL, PHYSICAL, STRUCTURAL AND SENSORY CHARACTERISTICS

María Dolores Alvarez[1,*], Wenceslao Canet[1], María José Jiménez[2] and Cristina Fernández[3]

[1]Department of Characterization, Quality and Safety, Institute of Food Science, Technology and Nutrition (ICTAN-CSIC), Ciudad Universitaria, Madrid, Spain

[2]USAS, Service Unit of Sensory Analysis, Institute of Food Science, Technology and Nutrition (ICTAN-CSIC), Ciudad Universitaria, Madrid, Spain

[3]Department of Food Science and Technology, Macromolecules Innovation Institute, Virginia Tech, US

---

[*] Corresponding Author Email: mayoyes@ictan.csic.es.

ABSTRACT

Growing awareness of the link between diet and health is fast changing consumer habits, so that there has been increasing demand for foods with health-enhancing properties. Extra virgin olive oil (EVOO) has important nutritional characteristics linked to its biophenol content and has very important antioxidant properties. The aim of this chapter was to study the effect of the addition of EVOO on instrumental textural properties, sensory texture profile analysis (TPA), and microstructure of fresh and frozen/thawed mashed potatoes formulated without and with added cryoprotectants (kappa-carrageenan [κ-C] and xanthan gum [XG]). EVOO behaves as soft filler due to droplet aggregates, whereas addition of cryoprotectants led to more structured mashed potatoes (MP) thanks to the gelling properties of κ-C. Both the percentage of added EVOO and processing had a much less significant effect on the texture of the MP containing κ -C and XG, evidencing the ability of this biopolymer blend to impart freeze/thaw stability. All samples with added EVOO were perceived as significantly softer and creamier than the samples without EVOO, whereas all MP samples with added cryoprotectants were perceived as significantly thicker and creamier than those without hydrocolloids. The results have shown that although instrumental textural data were able to explain differences in consistency perceived, structural information is needed to understand differences in creaminess. Back extrusion test is recommended to industry as practical quality control tool in the commercial production of MP with added EVOO.

Keywords: extra virgin olive oil; freezing; mashed potatoes; microstructure; texture profile analysis (TPA), sensory attributes

INTRODUCTION

Various health organizations recommend a daily intake of around 600 g of fruit and vegetables, but few people manage to consume this amount. Led by consumer demand, the food industry has shown an increased interest in the manufacture of healthier and more natural fruit and vegetable food products, such as soups, drinks and sauces [1]. Mashed potatoes (MP) made from 100% fresh potato tubers are in addition a natural vegetable semisolid food, which may also be suitable for freezing as a ready-meal component or as a product in itself such as potato gratin [2].

Olive oil is an important component of the diet of the countries surrounding the Mediterranean Sea. Due to its composition, olive oil is a good source of biophenols [3] as well as lipid-soluble and water-soluble vitamins (tocopherols, β-carotene, and ascorbic acid). In addition, thanks to its balanced fatty acid composition virgin olive oil has highly appreciated nutritional characteristics [4], known for a long time to the people of the Mediterranean region, who use it daily for a variety of culinary purposes. Biophenols with important antioxidant properties and a role in atherogenesis and cancer have been found and quantified in virgin and extra-virgin olive oils [5]. However, consumption has also increased in non-Mediterranean areas thanks to growing interest in the Mediterranean diet and a belief that it prevents certain diseases [3, 6]. A classic white sauce usually contains flour, milk and butter, but olive oil has been added to a white model-sauce to produce an innovative sauce approximating "Mediterranean cooking" [7].

The oil volume fraction exerts profound effects on the physicochemical and viscoelastic properties of emulsions, such as droplet size distribution, creaming, oxidative stability, and rheology [8]. Fat droplets influence the overall physicochemical and sensory properties of foods in a variety of different ways [9]. A great deal of research has been done on the influence of fat droplets on the rheology, stability and flavour of food emulsions, but less is known about their influence on emulsion appearance. Color is one of the major attributes affecting consumer perception of the quality of virgin olive oil [10], and chloroplast pigments (chlorophyll and carotenoids) are mainly responsible for the color of virgin olive oil, ranging from yellow–green to greenish gold [11].

Texture is by far the most important quality criterion for consumer sensory acceptance of freshly prepared and processed potato products, and particularly of frozen/thawed and dehydrated mashed potatoes. A fluffy and medium-consistency texture is desirable, whereas pastiness, gumminess and stickiness are negative attributes [12]. Texture instability remains the most significant challenge for frozen food products, especially with inevitable post-production temperature fluctuations. Loss of moisture and changes in textural attributes often result in significant reduction of product quality.

Previous studies showed that the addition of κ-C and XG to MP at a low concentration (each cryoprotectant at 1.5 g/kg) is recommendable on the basis of overall acceptability, especially when the product is going to be frozen [2, 13]. κ-C provides the appropriate texture, while XG imparts creaminess and mouthfeel to the product.

Few research has been done on the addition of olive oil in fresh and frozen/thawed mashed potatoes (designated FMP and F/TMP respectively), particularly with EVOO. The use of EVOO rather than commercial olive oil is preferable because of its high concentrations of both unsaturated fatty acids and antioxidant compounds such as polyphenols and tocopherols [14]. The purpose of the present research was to evaluate the effects of adding EVOO on the textural, physical, structural and sensory characteristics of FMP and F/TMP formulated without and with added cryoprotectants.

Materials and Methods

Materials

The potatoes used were fresh tubers (cv Kennebec) from Aguilar de Campoo (Burgos, Spain). κ-C (GENULACTA carrageenan type LP-60) and XG (Keltrol F [E]) were donated by Premium Ingredients, S.L. (Girona, Spain). EVOO (Carbonell, Spain) was chosen for addition to the MP. Following range-finding experiments, the lower and upper levels of EVOO to be used were set at 10 and 50 g/kg, respectively. A sample without EVOO was also prepared for each type of MP and processing conditions.

Preparation of MP Samples

Tubers were manually washed, peeled and diced. MP were prepared in ~ 2000-g batches from 607.7 g/kg of potatoes, 230.8 g/kg of semi-skimmed in-bottle sterilized milk (fat content, 15.5 g/kg), 153.8 g/kg of water, 7.7 g/kg of salt (NaCl) and the corresponding EVOO concentration (0, 10, 25, and 50 g/kg) using a TM 31 food processor (Vorwerk España, M.S.L., S.C., Madrid, Spain). MP were prepared without and with added κ-C and XG (MPA and MPB samples, respectively). In the latter case, hydrocolloids (each at 1.5 g/kg) were also added to the rest of the ingredients in the form of a dry powder. All the ingredients were cooked for 35 min at 90°C (blade speed: 40 rpm) [2, 13]. The mash was ground for 40 s (1,200 rpm) and for 20 s (2,600 rpm). The product was at once homogenized through a stainless steel sieve (diameter: 1.5 mm). The highest EVOO concentration was added twice to the MP to evaluate the effect of order of addition and EVOO thermal treatment on MP quality. First, 50 g/kg of EVOO was added along with the rest of the ingredients as indicated above, whereas in the second case the same EVOO concentration (designated "50b" g/kg) was added to the MP before final homogenization. Half of each fresh blend (FMP samples) was analysed immediately and the other half was frozen and thawed (F/TMP samples). Two repetitions of each composition were prepared in different weeks.

Freezing, Thawing and Heating Procedures

MP samples were placed on flat freezing and microwave thawing trays, and then frozen by forced convection with liquid nitrogen vapour in an Instron programmable chamber (model 3119-05, −70/+250°C) at −60°C until their thermal centres reached −24°C [13]. After freezing, the samples were packed in polyethylene plastic bags, sealed under light vacuum (−0.05 MPa) on a Multivac packing machine (Sepp Haggenmüller KG, Wolfertschwenden, Germany), and placed in a domestic freezer for storage at −24°C. Packed frozen samples were thawed in a Samsung M1712N microwave oven (Samsung Electronics S.A., Madrid, Spain). Samples were heated for 20 min at an output power rating of 600 W. Samples were brought up to 55°C by placing them in a Hetofrig CB60VS water bath (Heto Lab Equipment A/S, Birker∅d, Denmark). Sample testing was 55°C, where water and product temperatures were monitored by T-type thermocouples as described elsewhere [2, 15-17].

Instrumental Texture Measurements

Back extrusion (BE) and cone penetration (CP) mechanical tests were performed in order to study the empirical rheological behavior of "semisolid like" samples. Both experiments were performed using a TA.HDPlus Texture Analyser (Stable Micro Systems Ltd, Godalming, UK) equipped with a 300 N load cell. During tests, MP samples were kept at 55°C by means of a Temperature Controlled Peltier Cabinet (XT/PC) coupled to a separate heat exchanger and proportional-integral-derivative control unit. For performance of BE tests, a rig (model A/BE, Stable Micro Systems) was used consisting of a flat 45 mm diameter perspex disc plunger that was driven into a larger perspex cylinder sample holder (50 mm diameter) to force down into the MP samples and flow it upward through the concentric annular space between plunger and the container. The measuring cup was filled with 50 ± 1 g of MP. Product was extruded to a distance of 20 mm at 2 mm/s compression rate. At this point (most likely to be the maximum force), the probe returns to its original position. From the recorded force time curves, texture parameters with physical meaning are calculated, which vary from simple consistency indices to a derived flow behavior index, which is obtained according to the mathematical model suggested by Osorio and Steffe [18]. In this chapter, maximum positive force of extrusion (firmness (N)) and the negative area of extrusion (viscosity index (N s)) have been taken into account in order to describe texture changing in MP samples. For performing the CP tests, a TTC spreadability rig (HDP/SR, Stable Micro Systems) was used, consisting of a 45° conical perspex probe (P/45C) that penetrated a conical sample holder containing 7 ± 0.1 g of MP product. Product was penetrated to a distance of 17.5 mm at 3 mm/s compression rate. CP work per displaced volume (J/m^3) required to accomplish penetration was calculated from the area under the curve up to the "peak" or maximum penetration force, and the average force of the complete curve (N) was also recorded. Texture measurements were performed in quadruplicate and results averaged.

Other Quality Parameters

The color of the MP in the pots was measured with a HunterLab model D25 (Reston, VA, USA) color difference meter fitted with a 5 cm diameter aperture. Results were expressed in accordance with the CIELAB system with reference to illuminant D65 and a visual angle of 10°. The parameters determined were L^*, a^* and b^*. A higher L^* value indicated a brighter or whiter sample and values of a^* and b^* indicated red-green and yellow-blue colors. Yellowness index (YI) was calculated as $142.86b^*/L^*$ [17].

Expressible water (E_w) was measured by centrifugal force. Centrifuge tubes containing approximately 10 g of MP were centrifuged at 15,000×g for 30 min in a Sorvall®, RC-5B apparatus (Global Medical Instrumentation, Inc, Clearwater, Minnesota, USA). E_w was expressed as the percentage of liquid separated per total weight of sample in the centrifuge tube [19]. Measurements of color and E_w were performed in quadruplicate and the results averaged.

Sensory Analyses

MP samples were subjected to texture profile analysis (TPA) modified to evaluate vegetables purees according to UNE 87025 (1996) [20], which was used to select and define the sensory attributes included in the profile. A panel of 4 assessors, previously trained according to the ISO guidelines (ISO 8586-1:1993) [21] and with specific exercise in MP for 8 years [15-

17], evaluated the textural attributes of the samples. Profile attributes were classified into four groups [16]. Attributes are listed in the order of the perception according to ISO guidelines (ISO 13299: 2003) [22]: attributes perceived before putting the sample in mouth (granularity and moisture (1)); attributes perceived at the time of putting the sample in the mouth (stickiness, denseness, homogeneity, moisture (2) and firmness); attributes perceived at the time of preparing the sample in the mouth for swallowing (cohesiveness, adhesiveness and fibrousness (1)); attributes perceived during final and residual phases of mastication (ease of swallowing, palate coating and fibrousness (2)). A description of the sensory attributes evaluated during the TPA can be found elsewhere [16].

Samples were evaluated, in duplicate, in morning sessions (11:00 a.m.- 1:00 p.m.). Daily for 40 days assessors were given four samples (about 20 g each), for scoring attributes of each group in the texture profile. All the samples were served at 55 ± 1°C on Petri dishes. This sample temperature was reached and kept constant by placing the product in the Hetofrig CB60VS water bath prior to testing. For each sample, panelists evaluated the perceived intensity of the 13 attributes on 8 cm descriptive linear scales labelled at each anchor: (left anchor: 1 = ‘‘not detectable; right anchor: 9 = ‘‘extremely intense’’). To reduce fatigue a rest period of 5 min was taken after scoring each sample.

MP samples were also subjected to an overall acceptability (OA) test based on all sensory attributes (texture, color, taste) on a 9-point hedonic scale (with 8 cm) labelled at each anchor: (left anchor: 1 = “dislike extremely”; right anchor: 9 = “like extremely”). In this case, sensory assessment was conducted by a 14-member untrained panel. Every day, one sample (about 20 g each) was served under the same conditions as indicated above.

Scanning Electron Microscopy (SEM)

MP microstructure was examined by SEM using a Hitachi model S-2.100 microscope (CENIM-CSIC). MP samples were air-dried, then mounted and sputter-coated with Au (200 Å aprox.) in a SPI diode sputtering system metallizer. Photomicrographs were taken with a digital system Scanvision 1.2 of RONTEC (800x1.200 pixel).

Statistical Analysis

A three-way ANOVA with interactions was applied to evaluate how the three factors studied—EVOO concentration, presence or absence of hydrocolloids and performance or not of processing —affected the texture, color, sensory attributes and the OA of the MP. E_w was always zero for the MPB samples; a two-way ANOVA with interactions was applied to evaluate how EVOO concentration and processing affected the E_w of the products. Minimum significant differences were calculated using Fisher's least significant difference test (LSD, 99% for comparison of instrumental parameters and 95% for comparison of sensory attributes and OA). Analyses were performed with Statgraphics® software version 5.0 (STSC Inc., Rockville, MD, USA).

Results and Discussion

Instrumental Texture Measurements

Table 1 shows the effects of EVOO concentration, cryoprotectant addition and processing on the values of the textural properties derived from the BE and CP tests. Samples with added κ-C and XG, as well as those subjected to freezing/thawing, presented significantly higher and lower textural properties than their respective counterparts. Previous studies showed that when κ-C/XG blends were added to FMP and F/TMP samples, κ-C provided the appropriate texture whereas XG imparted creaminess to the product [2, 13, 23]. Analogously, in starch/XG blends, it was observed that XG does not interfere in potato starch network building

[7, 24]. Therefore, addition of both hydrocolloids to MP produces a more structured system which is associated with the gelling properties of κ-C. In natural MP, the product was softer than the fresh control after freezing and thawing [15]. MP is a starchy food, and as such may present quality problems such as syneresis and organoleptic and textural changes. These problems have been ascribed to phase separation caused by retrogradation of the starch [19, 25].

Table 1. Effects of EVOO concentration, cryoprotectant addition and freezing/thawing on textural properties of MP

Source	Firmness (N)	Viscosity index (N s)	Work per displaced volume (J/m^3)	Average force (N)
Main effects:				
A: EVOO concentration (g/kg)				
0	6.21 a	-29.33 a	3518.78 a	1.51 a
10	6.08 a	-28.37 a	3462.25 a	1.49 a
25	5.32 b	-26.12 b	2867.16 b	1.23 b
50	4.69 c	-23.06 c	2681.10 b	1.15 b
50b	4.74 c	-23.69 c	2786.29 b	1.19 b
P values	<0.001	<0.001	<0.001	<0.001
LSD (99%)	0.26	1.24	227.00	0.097
B: Cryoprotectant addition				
Without κ-C and XG	4.71 a	-20.71 a	2333.51 a	1.00 a
With κ-C and XG	6.10 b	-31.52 b	3792.72 b	1.63 b
P values	<0.001	<0.001	<0.001	<0.001
LSD (99%)	0.16	0.78	143.57	0.06
C: Processing				
Fresh	5.62 a	-26.73 a	3248.00 a	1.39 a
Frozen/thawed	5.19 b	-25.50 b	2878.23 b	1.23 b
P values	<0.001	<0.001	<0.001	<0.001
LSD (99%)	0.16	0.78	143.57	0.06
Interactions				
AB	<0.001	<0.001	<0.001	<0.001
AC	0.001	0.18	<0.001	<0.001
BC	<0.001	<0.001	<0.001	<0.001
ABC	<0.001	<0.001	<0.001	<0.001

MP, mashed potato; EVOO: extra virgin olive oil; κ-C: kappa-carrageenan; XG: xanthan gum; LSD, least significant difference.

With respect to the effect of EVOO addition, the maximum textural property values were registered in the samples without EVOO, although differences between textural properties of samples with 10 g/kg added EVOO and those without EVOO were non-significant (Table 1). However, increasing EVOO concentration produced softer, liquid-like systems, indicating that EVOO behave as soft filler. This result is to be expected as increasing concentrations of liquid oil are added to the product, increasing the oil-phase volume fraction. In oil-in-water emulsions, the extent of the linear region decreased with increasing oil-phase volume fraction from 20 to 40% v/v [26]. For their part, Dickinson and Chen [8] suggested that oil/water emulsions may undergo a behavior transition from predominantly entropic behavior to predominantly enthalpic behavior, with increasing oil-phase volume fraction.

The ANOVA also showed that the three binary interactions had a significant effect on instrumental firmness, work per displaced volume and average force (Table 1). This means that the effect of EVOO concentration on the texture depended on the presence of κ-C and XG, and on the freezing/thawing of the systems. Besides, EVOO concentration and cryoprotectant addition (AB) and cryoprotectant addition and processing (BC) interactions also significantly affected the viscosity index from the BE tests.

From the variation in the firmness value based on EVOO concentration for both MPA and MPB samples shown in Figure 1a, one can observe that firmness was lower in the MPA than in the MPB samples; moreover, the variation in sample firmness was much greater when EVOO content increased from 10 to 50 g/kg in the MPA samples than in the MPB ones. Also, when the concentration of added EVOO was increased, the firmness value behaved similarly in the FMP and F/TMP samples (Figure 1b); in both cases, the increase in EVOO content led to reduced firmness, without important differences between 50 and 50b g/kg. As droplet concentration increases, the droplets are polydispersed and the samples present a less close packing structure. In mayonnaise, increasing walnut oil content increases the diameter of oil droplets and consequently reduces viscoelastic properties [27].

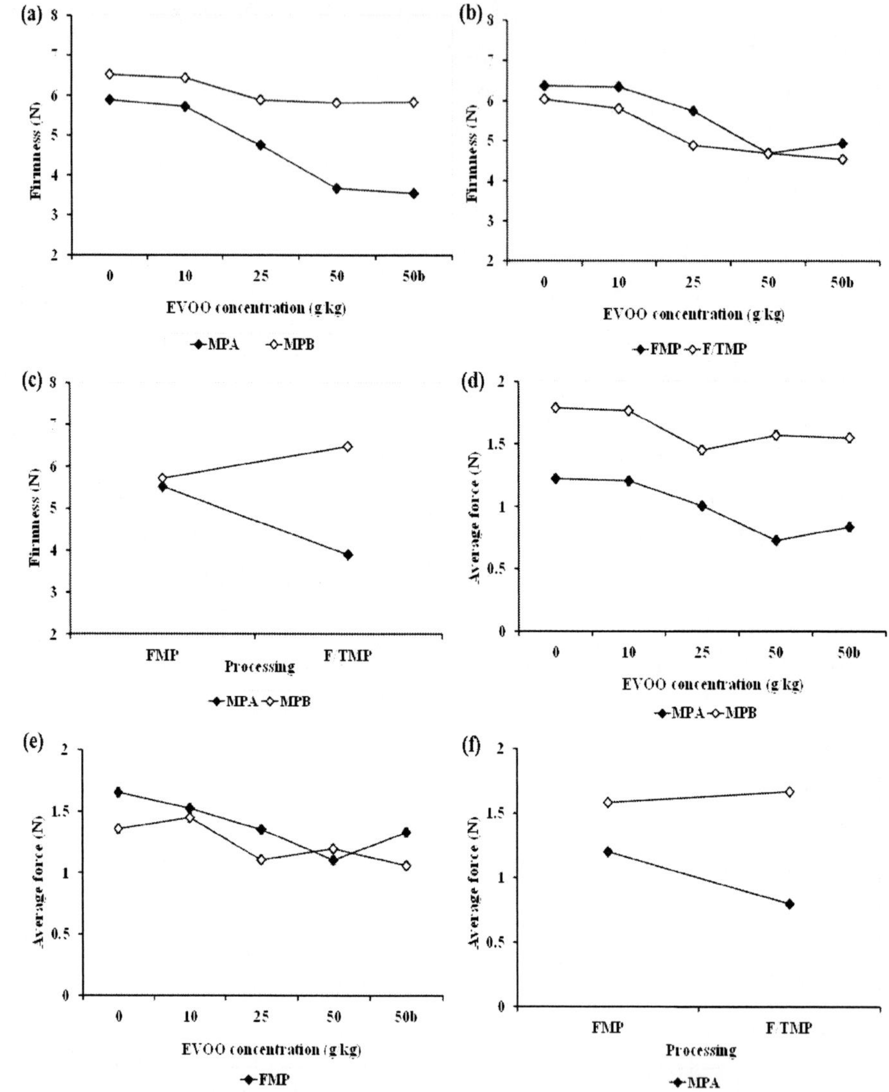

Figure 1. Textural properties of mashed potatoes with added extra virgin olive oil (EVOO). (a–c) Firmness; (d–f) Average force; MPA, MPB: mashed potatoes without and with added cryoprotectants, respectively; FMP, F/TMP: fresh and frozen/thawed mashed potatoes, respectively.

From the variation in the firmness based on processing, the firmness value developed differently for the MPA and MPB samples (Figure 1c).

Processing significantly reduced sample firmness in the MPA samples but significantly increased it in MPB samples. This behavior can be explained, taking account that much stronger and more cohesive networks are formed when solutions of XG are frozen and thawed [28]. The effect of XG may be explained by amylose/XG interactions, which compete against amylose/amylose interactions, retarding or even preventing retrogradation. Also, the addition of small amounts of XG to white sauces made with starches from different sources significantly improves freeze/thaw stability [29].

In turn, the variation in average force with EVOO concentration or both MPA and MPB samples (Figure 1d) was similar to that observed in firmness. In this case of the MPB samples, the ones with 25 g/kg EVOO added had poorer consistency, whereas in the MPA systems, the ones with 50 g/kg had poorer consistency. When the EVOO concentration was increased, the average force decreased in both FMP and F/TMP samples (Figure 1e), although in the latter case adding 10 g/kg EVOO slightly increased the average force as compared with the control without EVOO. Both the BE firmness and the CP average force values were greater when the EVOO was added after cooking (50b g/kg) in the FMP samples but not in the F/TMP samples. When the processing-dependent variation in average force was plotted (Figure 1f), the changes in that value were also similar to those observed for firmness (Figure 1c). Plots for the viscosity index and the work per displaced volume have been omitted.

Color Measurements and Expressible Water

All of the three factors studied significantly changed the color parameters, although processing did not significantly affect the YI (Table 2). An increase in EVOO level favors higher L^* value (lightness) due to an increase in the overall light scattering associated with the scattering properties of fat [9]. As the EVOO concentration increased, there was an increase in redness (decreasingly negative a^* values) and in yellowness (YI), associated with the augmented pigment content of the MP. The pigment profile of the virgin olive oil comprises chlorophyll a, chlorophyll b and derivative pigments associated with the acidic medium of the oil extraction process [10].

Table 2. Effects of EVOO concentration, cryoprotectant addition and freezing/thawing on color measurements and expressible water of MP

Source	*L**	*a**	YI	E_w (%)
Main effects:				
A: EVOO concentration (g/kg)				
0	60.79 a	-3.85 a	10.14 a	22.42 b
10	62.95 b	-3.70 b	13.73 b	25.73 a
25	63.68 c	-3.54 c	18.58 c	20.72 d
50	66.04 d	-3.11 d	24.08 d	21.52 c
50b	65.70 e	-3.14 d	23.47 e	21.85 b, c
P values	<0.001	<0.001	<0.001	<0.001
LSD (99%)	0.07	0.02	0.20	0.71
B: Cryoprotectant addition				
Without κ-C and XG	62.85 a	-3.59 a	16.74 a	-
With κ-C and XG	64.81 b	-3.34 b	19.26 b	-
P values	<0.001	<0.001	<0.001	-
LSD (99%)	0.04	0.01	0.13	-
C: Processing				
Fresh	63.33 a	-3.53 a	18.03 a	21.01 a
Frozen/thawed	64.33 b	-3.40 b	17.97 a	23.89 b
P values	<0.001	<0.001	0.17	<0.001
LSD (99%)	0.04	0.01	0.13	0.45
Interactions				
AB	<0.001	<0.001	<0.001	-
AC	<0.001	<0.001	<0.001	<0.001
BC	<0.001	0.34	<0.001	-
ABC	<0.001	<0.001	<0.001	-

MP, mashed potato; EVOO: extra virgin olive oil; κ-C: kappa-carrageenan; XG: xanthan gum; L*: lightness (L* = 0 [black], L* = 100 [white]), a*: color position between green and red (-a* = green, +a* = red); YI: yellowness index; LSD, least significant difference.

L^* increased when *k*-C and XG were added to the MP, which could be partially due to their absolute water-holding capacity (WHC) as discussed below. Also, a^* was higher in the MPB than in the MPA samples, indicating significant raise sample redness. The loss of greenness associated with cryoprotectant addition was probably due to the presence of XG in the system as found previously [17].

Increased lightness in the F/TMP samples as compared with their FTM counterparts may have been partly due to the formation of fissures produced by the growth of ice crystals during freezing, which favors the release of water; this would transmit the light more rather than capturing it. For its part, the loss of greenness found in the processed samples (a^* values nearer to 0) as compared with the fresh counterparts could be due to slight non-enzymatic browning (Maillard reaction) during microwave thawing.

On the other hand, the three interactions had a significant effect on L^* and YI (Table 2). Moreover, AB and EVOO concentration and processing (AC) interactions significantly affected the a^* value. The variation in the L^* value based on EVOO concentration in both MPA and MPB samples (Figure 2a) shows that increased EVOO concentration produced an increase in the L^* value in both samples. The influence of droplet characteristics on the optical properties of colored oil-in-water emulsions has been studied [9]. The lightness of the emulsions increased with increasing droplet concentration and decreasing droplet size. As the droplet concentration increases, so does the reflectance because the droplets scatter light more effectively and hence the light beam is unable to penetrate further into the product and be absorbed.

The differences between the L^* values of the MPA samples and their MPB counterparts increased with increasing the EVOO content (Figure 2a). In emulsions, XG is added to the aqueous phase to prevent droplets from rapidly creaming and coalescence [26, 30].

In this chapter, oil droplet diameters were not measured, but it is probable that the droplets in the MPB samples were smaller than in the ones without cryoprotectants, as the presence of XG in the system would prevent coalescence. The reason why the L^* values were lower in the MPA samples, then, is that reflectance decreases with increasing droplet diameter. Note that in the MPB samples, the L^* value was greater when the EVOO was added after cooking (50b g/kg), whereas in the MPA systems it was greater in the samples with 50 g/kg EVOO added before cooking.

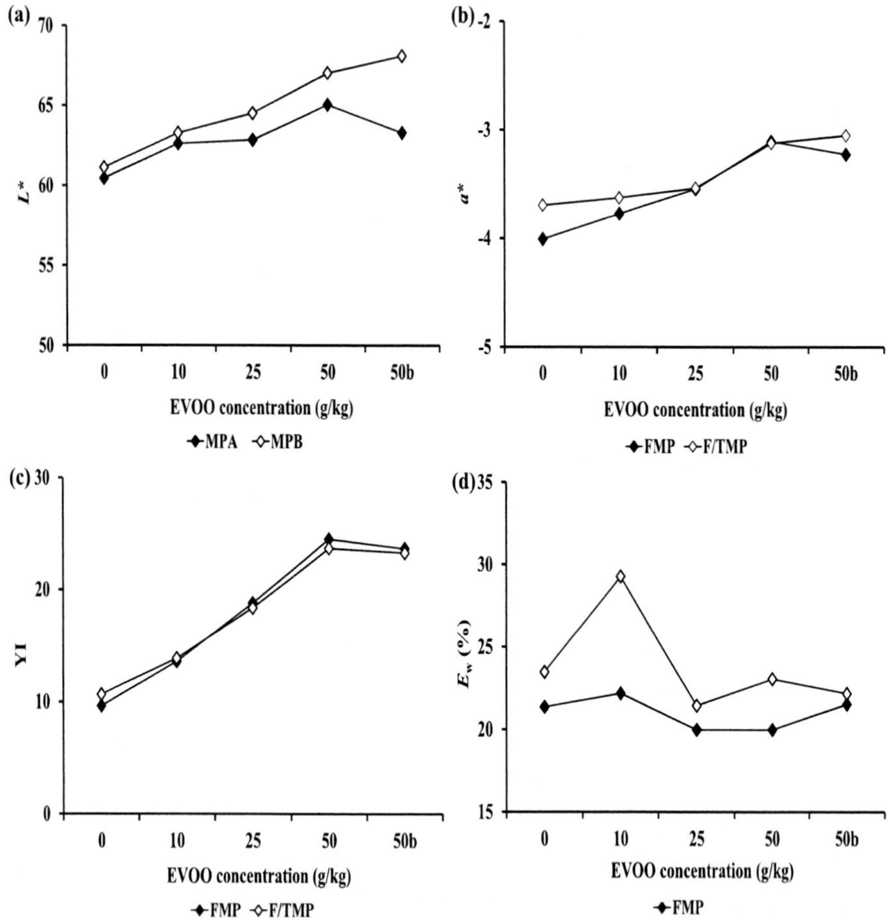

Figure 2. Color parameters and expressible water of mashed potatoes with added extra virgin olive oil (EVOO). (a–c) L^*, lightness; a^*, red-greenness; YI, yellowness index; (d) E_w, expressible water; MPA, MPB: mashed potatoes without and with added cryoprotectants, respectively; FMP, F/TMP: fresh and frozen/thawed mashed potatoes, respectively.

This discrepancy could also be related to the presence of cryoprotectants in the system. MP with EVOO added before final homogenization would be expected to have larger droplets because the oil was not thoroughly triturated. In the presence of XG, the droplets scatter light more effectively when the oil is not so strongly entrapped in the matrix. In the MPA samples on the other hand, reflectance probably decreased because the scattering efficiency of the droplets decreases above a certain droplet size [9].

In turn, as the droplet concentration increases, more reflected light travels through the oil phase of the MP being absorbed by the pigments mentioned earlier, intensifying the color of the MP (Figures 2b, 2c). However, as regards to YI values, there were small differences between FMP and F/TMP samples. Anyway, the color differences found between samples, although significant, should not be of major importance in practical terms.

E_w changed significantly with EVOO concentration and processing (Table 2), and the AC interaction had a significant effect on the WHC of the samples (Figure 2d). In this chapter, addition of κ-C and XG reduced the E_w of both FMP and F/TMP samples to 0%, corroborating the well-established ability of XG to reduce water separation [2, 16, 29], and evidencing the existence of XG-water or XG-water-XG interactions in the systems. XG is an anionic, hygroscopic material of exceptional pseudoplasticity [31]; its texturizing effect can be achieved at low gum concentration because of unusual water holding ability. Also, adding XG (0.3% w/w) to corn starch pastes (10% w/w) minimized amylose retrogradation, syneresis and rheological changes after freezing [32]. Certainly, the E_w values confirm that XG effectively stabilizes MP against syneresis when no more than 1.5 g/kg is added.

Besides, WHC was greater in the FMP samples than in their F/TMP counterparts at all EVOO concentrations (Figure 2d). This result is probably related to structural damage caused by freezing. The addition of EVOO at low concentrations significantly increased E_w, mainly in the processed samples, which is likely due to that the interchain spaces were occupied by oil, displacing the water [34]. However, the addition of EVOO at higher concentrations significantly reduced water loss, probably because excess oil hindered the release of water from the starch matrix. EVOO by itself was not effective in enhancing the WHC of MP. In any case E_w percentages were also quite high (>20) in both FMP and F/TMP samples without added EVOO, evidencing the presence of weak starch–water or starch–water–starch interactions in all the systems. Water separation in the MPA samples is related to starch retrogradation and consequent reduction of WHC [33].

Sensory Analyses

Attributes Perceived before Putting the Sample in Mouth

All the three main factors and their interactions significantly ($P < 0.05$) affected the scores for granularity and moisture (1) (Table 3). One can observe that at all EVOO concentrations, granularity scores were greater in the MPA samples (Figure 3a) and likewise in the fresh products (Figure 3b). Christianson et al. [35] indicated that gums like XG affect the gelatinization and retrogradation of starch through strong associations with amylose, resulting in reduced amylose–amylose interactions. In turn, presence of XG reduced granularity in the F/TMP systems by assisting new starch/water interactions and consequent water absorption. In both MPA and MPB samples, panelists judged granularity lowest in the samples with more than 10 g/kg added EVOO. The effects of EVOO on granularity are related to the lubricating and coating properties conferred by the oil as reported for vanilla custard desserts [36].

In turn, moisture (1) decreased significantly with respect to MPA samples with the addition of cryoprotectants in both FMP and F/TMP (Figure 3c). Panelists detected greater ability to hold water molecules in MPB samples, confirming the results for E_w values. Similarly, panelists detected less aqueousness in the processed samples than in the fresh ones, probably due to water loss.

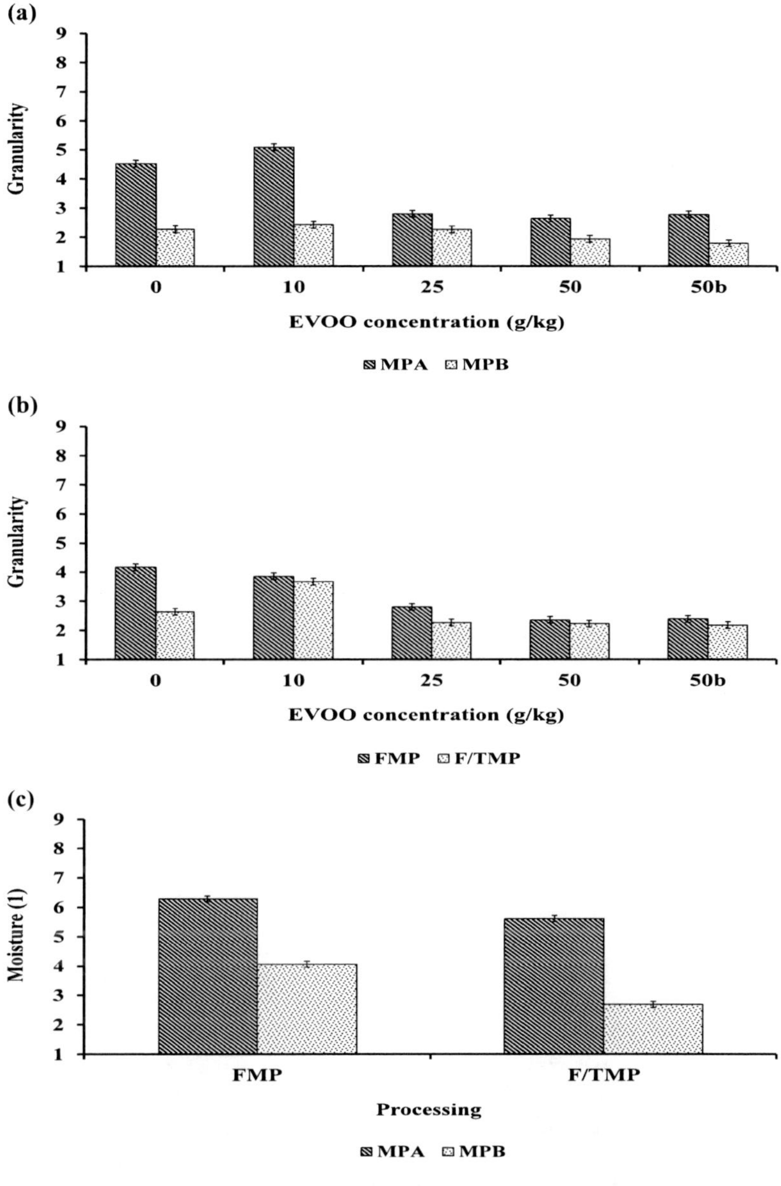

Figure 3. Texture profile analysis sensory attributes of mashed potatoes with added extra virgin olive oil (EVOO) perceived before of putting the sample in the mouth: (a, b) Granularity; (c) Moisture (1); MPA, MPB: mashed potatoes without and with added cryoprotectants, respectively; FMP, F/TMP: fresh and frozen/thawed mashed potatoes, respectively.

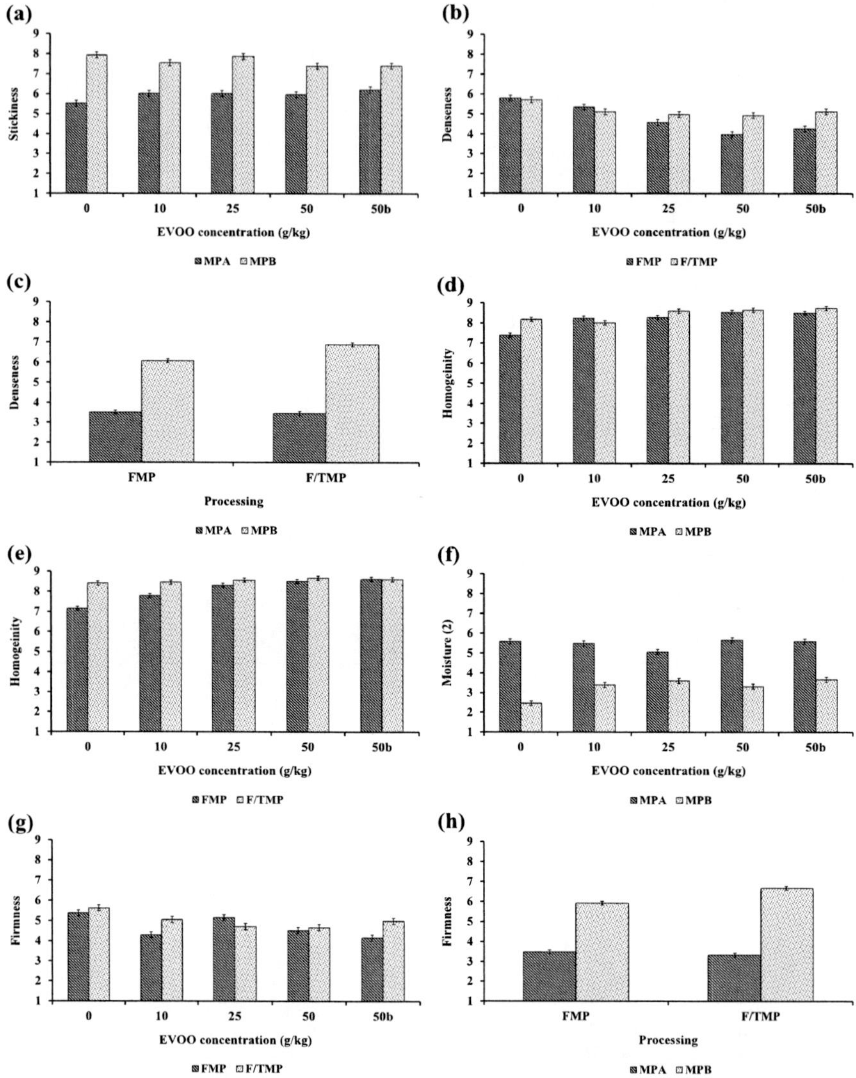

Figure 4. Texture profile analysis sensory attributes of mashed potatoes with added extra virgin olive oil (EVOO) perceived at the time of putting the sample in the mouth: (a) Stickiness; (b, c) Denseness; (d, e) Homogeneity; (f) Moisture (2); (g, h) Firmness; MPA, MPB: mashed potatoes without and with added cryoprotectants, respectively; FMP, F/TMP: fresh and frozen/thawed mashed potatoes, respectively.

Table 3. Effects of EVOO concentration, cryoprotectant addition and freezing/thawing on sensory attributes of MP perceived before and at the time of putting the sample in the mouth

Sensory attributes	Perceived before putting the sample in the mouth		Perceived at the time of putting the sample in the mouth				
Source	Granularity	Moisture (1)	Stickiness	Denseness	Homogeneity	Moisture (2)	Firmness
Main effects:							
A: EVOO concentration	<0.001	0.022	0.487	<0.001	<0.001	<0.001	<0.001
B: Cryoprotectant addition	<0.001	<0.001	<0.001	<0.001	<0.001	<0.001	<0.001
C: Processing	<0.001	<0.001	0.542	<0.001	<0.001	<0.001	0.002
Interactions							
AB	<0.001	0.002	0.002	0.236	<0.001	<0.001	0.044
AC	<0.001	0.003	<0.001	<0.001	<0.001	0.082	<0.001
BC	0.015	0.001	0.611	<0.001	0.970	0.003	<0.001
ABC	<0.001	<0.001	<0.001	0.143	<0.001	<0.001	<0.001

EVOO: extra virgin olive oil; MP: mashed potatoes.

Table 4. Effects of EVOO concentration, cryoprotectant addition and freezing/thawing on sensory attributes of MP perceived at the time of preparing the sample for swallowing and during the final and residual phases of mastication, and overall acceptability

Sensory attributes	Perceived at the time of preparing the sample for swallowing			Perceived during the final and residual phases of mastication			Overall acceptability (OA)
Source	Cohesiveness	Adhesiveness	Fibrousness (1)	Ease of swallowing	Palate coating	Fibrousness (2)	
Main effects:							
A: EVOO concentration	<0.001	<0.001	<0.001	<0.001	<0.001	<0.001	<0.001
B: Cryoprotectant addition	<0.001	<0.001	<0.001	<0.001	0.520	<0.001	<0.001
C: Processing	0.002	<0.001	<0.001	<0.001	0.601	<0.001	<0.001
Interactions							
AB	<0.001	<0.001	<0.001	<0.001	<0.001	<0.001	<0.001
AC	0.292	0.818	<0.001	0.003	<0.001	<0.001	<0.001
BC	0.083	<0.001	<0.001	0.126	<0.001	<0.001	0.055
ABC	<0.001	<0.001	<0.001	0.211	<0.001	<0.001	0.015

EVOO: extra virgin olive oil; MP: mashed potatoes.

Attributes Perceived at the Time of Putting the Sample in the Mouth

Stickiness scores were significantly higher in the MPB samples, although there were no differences in these scores as a consequence of EVOO concentration or processing (Table 3, Figure 4a). In turn, the three factors significantly affected scores for denseness, homogeneity, moisture (2) and firmness. Denseness was significantly higher in the processed samples than in the fresh samples only when EVOO was added at the highest concentrations (Figure 4b).

Also, denseness was lower in the MPA than in the MPB samples (Figure 4c), and only in this latter case were denseness scores significantly higher in the F/TMP samples than in their FMP counterparts. When EVOO concentration was increased, homogeneity increased in both MPA and MPB samples (Figure 4d). Note that the presence of EVOO in the systems rendered differences in homogeneity among MPA and MPB samples less appreciable. Also, when EVOO concentration was increased (Figure 4e), homogeneity increased in the FMP products, but was almost constant in the processed samples. This indicates a positive effect of adding EVOO to MP, as the negative effect of freezing on this attribute is masked by the oil. Panelists detected reduced moisture (2) in the MPB samples and in the processed systems, and when the EVOO concentration was increased, moisture (2) significantly increased when cryoprotectants were also added (Figure 4f).

In turn, panelists detected reduced firmness in the samples with added EVOO, without added cryoprotectants and without processing. One can observe that the processed samples with the lower and higher EVOO concentrations were the firmest, whereas in the systems with 25 g/kg added EVOO, the fresh samples had similar firmness than the control (Figure 4g). In the MPA samples, there were no differences between firmness scores in fresh and processed samples (Figure 4h); however, panelists detected increased firmness in processed MP with added κ-C and XG, matching the result for textural properties in MPB samples (Figures 1c, 1f).

Attributes Perceived at the Time of Preparing the Sample in the Mouth for Swallowing

EVOO concentration, cryoprotectant addition and processing also had a significant effect on cohesiveness, adhesiveness and fibrousness (1) (Table 4). When EVOO concentration was increased, cohesiveness and adhesiveness scores decreased significantly in the MPB samples (Figures 5a, 5b).

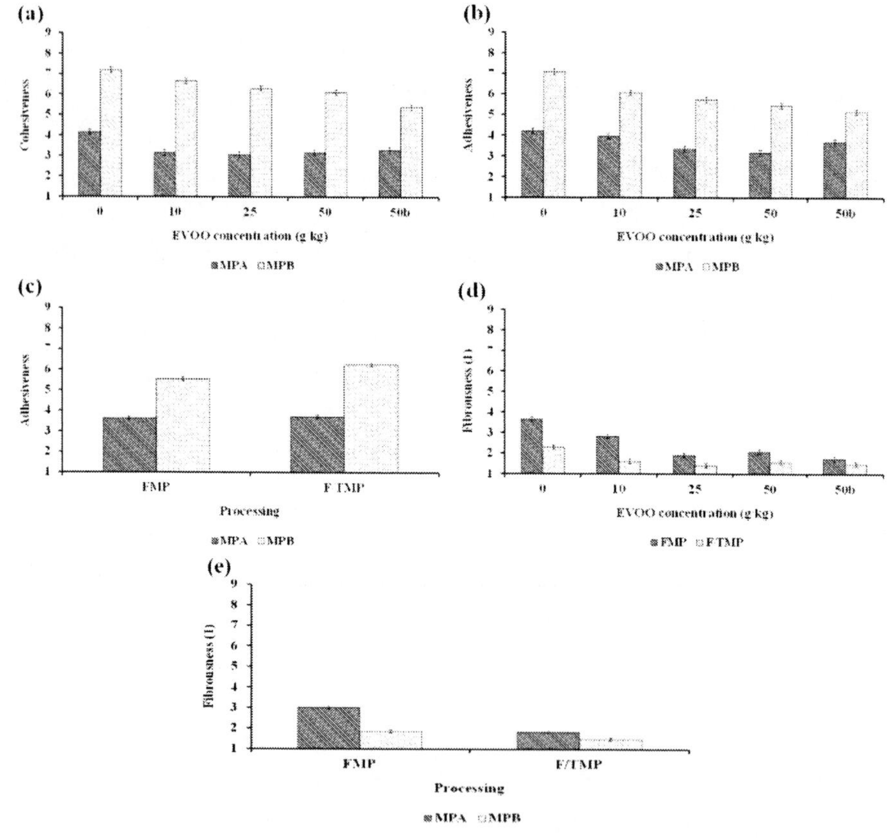

Figure 5. Texture profile analysis sensory attibutes of mashed potatoes with added extra virgin olive oil (EVOO) perceived at the time of preparing the sample for swallowing: (a) Cohesiveness; (b, c) Adhesiveness; (d, e) Fibrousness (1); MPA, MPB: mashed potatoes without and with added cryoprotectants, respectively; FMP, F/TMP: fresh and frozen/thawed mashed potatoes, respectively.

In the MPA samples, there were no significant differences between the adhesiveness scores of fresh and processed samples (Figure 5c), whereas panelists scored the processed MPB samples higher for adhesiveness than their fresh counterparts. Scores for fibrousness (1) also decreased with increasing EVOO concentration, with cryoprotectant addition and with processing (Figures 5d, 5e). Again, addition of cryprotectants reduced differences in fibrousness (1) between fresh and processed samples. This is probably related to the fact that the hydrocolloids can make systems in the rubbery state more viscous, reducing molecular mobility and preventing retrogradation [32].

Attributes Perceived during Final and Residual Phases of Mastication

The three factors studied had a significant effect on ease of swallowing and fibrousness (2) (Table 4), whereas only EVOO concentration had a significant effect on palate coating. In samples both without and with added cryoprotectants (Figure 6a) and in both FMP and F/TMP samples (Figure 6b), ease of swallowing scores increased with increasing oil content. However, only when EVOO was added at concentrations of 0 and 10 g/kg, the scores for this attribute were higher in the samples without cryoprotectants and in the processed samples. Panelists also scored the samples with added EVOO significantly higher for palate coating than the ones made without EVOO (Figures 6c, 6d).

Scores for palate coating were higher in the MPA samples with 25 and 50 g/kg added EVOO than in their MPB counterparts (Figure 6c), and the EVOO content had a much smaller effect in the F/TMP samples than in the fresh counterparts (Figure 6d). Palate coating scores for MPA samples decreased after processing, whereas scores for MPB samples increased with respect to the fresh products (Figure 6e). Also, in the MPA samples, the addition of EVOO at all concentrations significantly reduced sample fibrousness (2) (Figure 6f). A complete dependence study was performed on the instrumental textural properties versus sensory attribute scores. Low correlations between instrumental and sensory ratings were found. Previous publications by other researchers generally agree on good to excellent correlations for hardness (based on calculated "*r*" values) [37].

Correlations for other parameters are usually less good and product-dependent. In this chapter, relatively good correlations with sensory denseness and adhesiveness scores were found only in the case of viscosity index ($R^2 = 0.81$ and 0.76, respectively).

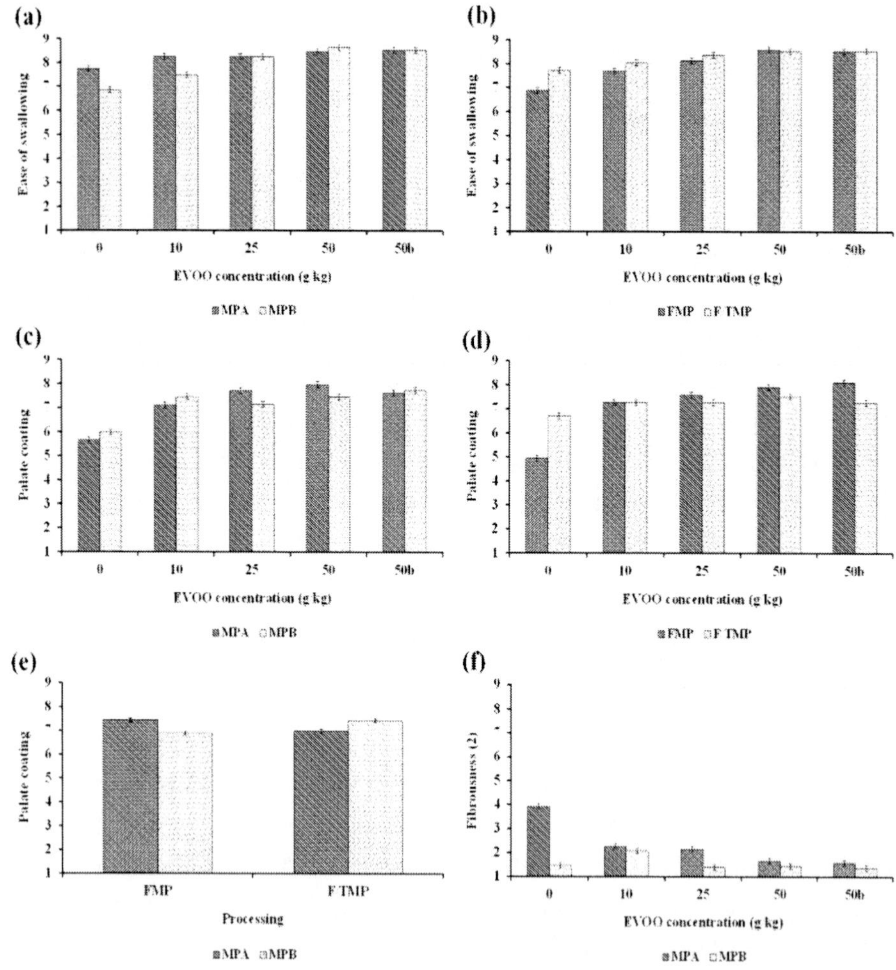

Figure 6. Texture profile analysis sensory attributes of mashed potatoes with added extra virgin olive oil (EVOO) perceived during the final and residual phases of mastication: (a, b) Ease of swallowing; (c-e) Palate coating; (f) Fibrousness (2); MPA, MPB: mashed potatoes without and with added cryoprotectants, respectively; FMP, F/TMP: fresh and frozen/thawed mashed potatoes, respectively.

Differences in consistency observed among samples were explained by viscosity index, but not the variation in granularity or fibrousness, determining the sample creaminess.

Overall Acceptability

EVOO concentration, cryoprotectant addition and processing had a significant effect on the OA of the samples (Table 4). Scores for OA increased significantly with increasing EVOO content in both MPA and MPB samples (Figure 7a), and likewise in both FMP and F/TMP samples (Figure 7b). Similarly, a positive relationship between oil content and sensory acceptability has been observed in a set of Polish commercial mayonnaises [38] and in salami [14]. In this chapter, the main differences between samples without and with added EVOO were ascribed to either an aromatic or a creamy note detected in the oil-added MP. Samples with higher percentages of EVOO produced less sensation of dryness and roughness, more sensations of flavor, creamy and fatty mouthfeel and after-feel than the samples without added oil. Fat is a well-known enhancer of creaminess sensations [36]. The latter authors suggested that the possible mechanism by which fat affects the sensory attributes include lubrication and flavor release. The effects of fat on odor and flavor attributes may be related to the flavor-releasing properties of fat.

Panelists scored the MPB and F/TMP samples higher for OA (Figures 7a, 7b). This is probably related to the presence of XG in the systems. It was found that samples containing blends of κ-C and XG [2, 13], were preferred organoleptically due to the creamy mouthfeel they produced. The effects of XG on mouth texture may be related to its WHC, as perceived by the panelists.

Besides, in the processed MPB samples, there were no significant differences between the OA scores given to the MP at any concentration of added EVOO (which were the highest). This has important consequences for the formulation of EVOO-based MP. Results indicate that in the presence of κ-C and XG, if the EVOO content is reduced to below 25 g/kg, the OA score for the product does not decrease, and hence its consumer acceptability is not adversely affected.

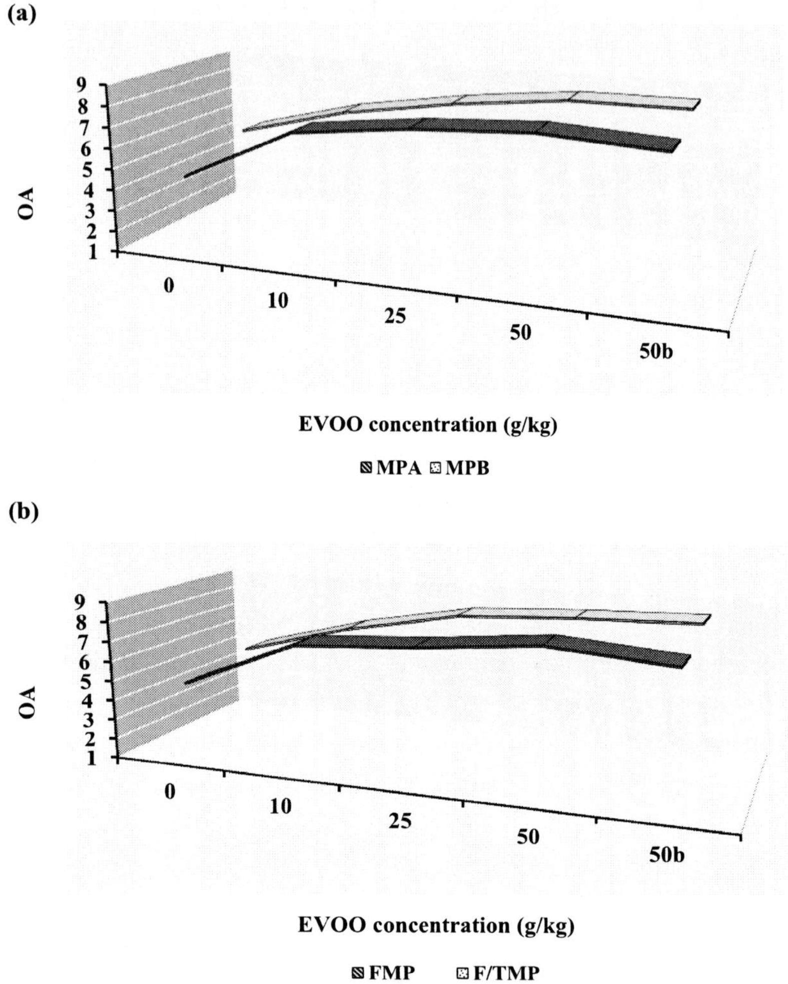

Figure 7. Overall acceptability of mashed potatoes with added extra virgin olive oil (EVOO); MPA, MPB: mashed potatoes without and with added cryoprotectants, respectively; FMP, F/TMP: fresh and frozen/thawed mashed potatoes, respectively.

Microstructure Examination of MP

To achieve a better understanding of the sensory and rheological results and the effect of adding cryoprotectants and of freezing/thawing, the microstructure of the MP samples was studied by SEM (Figures 8-11). Figure 8a shows a microphotograph of the fresh control without either added cryoprotectants or oil. Cooked cells are still distinguishable and firmly bound together by a continuous network of amylose. However, in the fresh control without added EVOO but with added cryoprotectants (Figure 8b), less complete cells are visible, appearing separated from one another and embedded in a continuous network of amylose and κ-C in which starch granules and XG aggregates are entrapped. Probably, the presence of cryoprotectants occluding a great amount of water probably facilitated loss of the original cell shape.

Microphotographs of the corresponding processed counterparts (Figures 9a, 9b) show that freezing and thawing of MPA and MPB samples resulted in completely dissolved cells. Part of the intracellular water was drawn out osmotically because of freezing-induced concentration of the cell mass. Cell tearing is probably caused by the formation of ice crystals. Fresh MPA sample contain more complete cells (Figure 9a), which could give them greater mechanical strength; this would justify that the values of the textural properties were higher in fresh MPA samples than in their processed counterparts. In turn, the processed MPA sample without added EVOO (Figure 9a) developed a spongy appearance due to amylose and amylopectin retrogradation occurring during freezing and frozen storage [32].

The microphotograph of processed MPB sample without added EVOO (Figure 9b) shows the presence of fibers or strands. According to Giannouli and Morris [28], during freezing, XG chains are forced to align and associate by conversion of water to ice crystals. The forced associations survive upon thawing to give a cryogel network. It is likely that such strands are related to this XG conformational transition, as they were observed in most of the F/TMP containing cryoprotectants. Formation of strands can be explained by a progressive increase in local concentration of the polymer as liquid water is converted into ice crystals, promoting intermolecular associations.

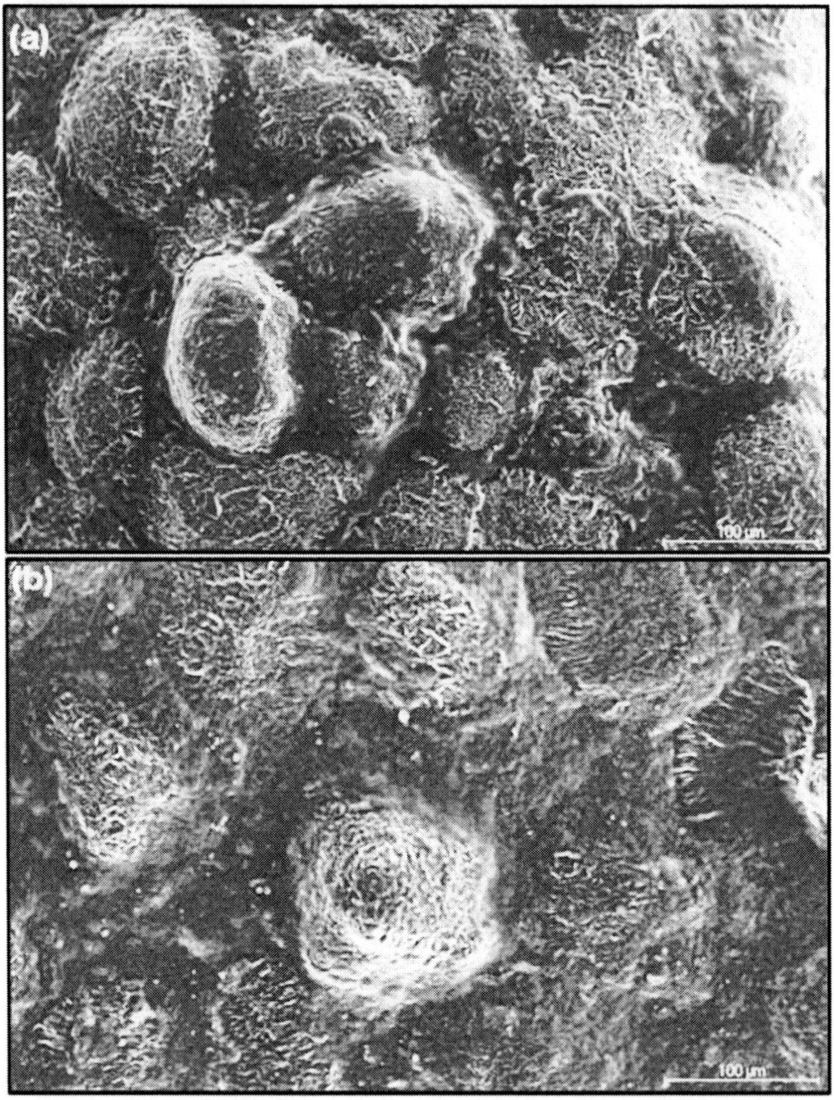

Figure 8. Microphotographs of mashed potatoes (a) Fresh sample without added cryoprotectants; (b) Fresh sample with added cryoprotectants; Magnification was 200 (bar = 100 μm).

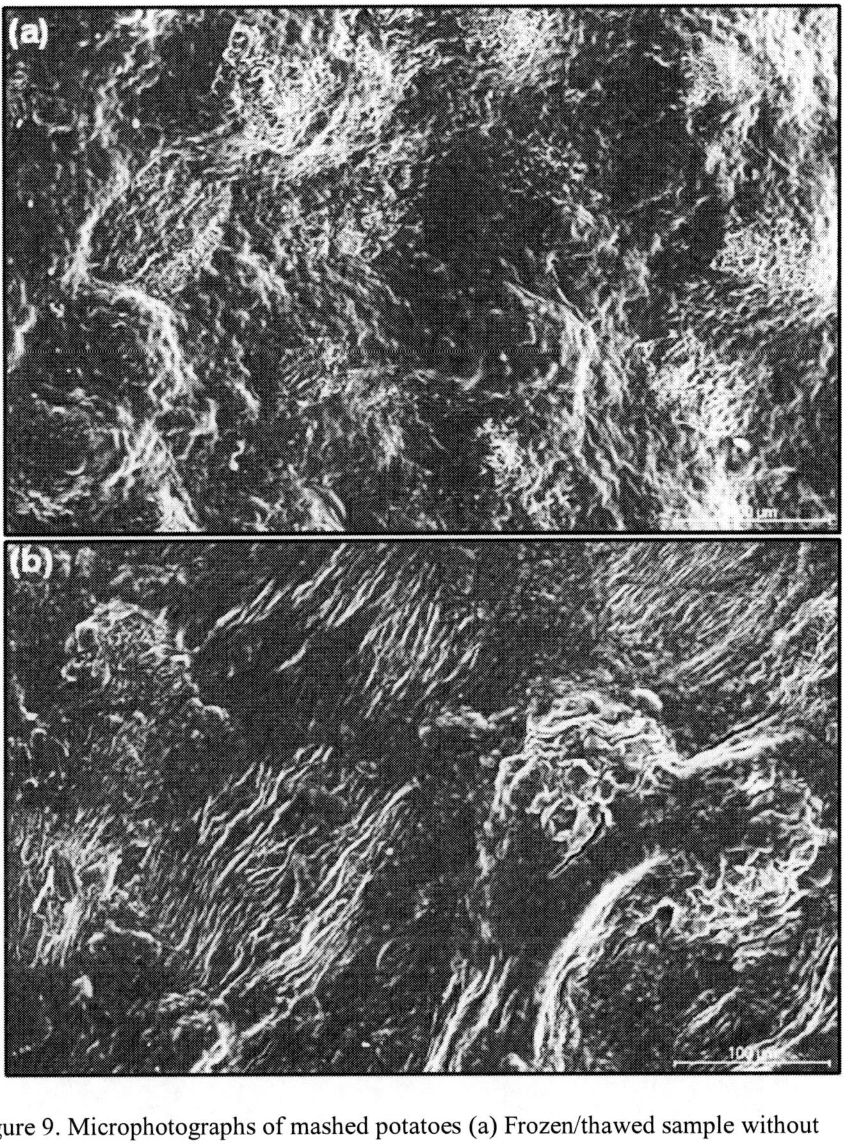

Figure 9. Microphotographs of mashed potatoes (a) Frozen/thawed sample without added cryoprotectants; (b) Frozen/thawed sample with added cryoprotectants; Magnification was 200 (bar = 100 μm).

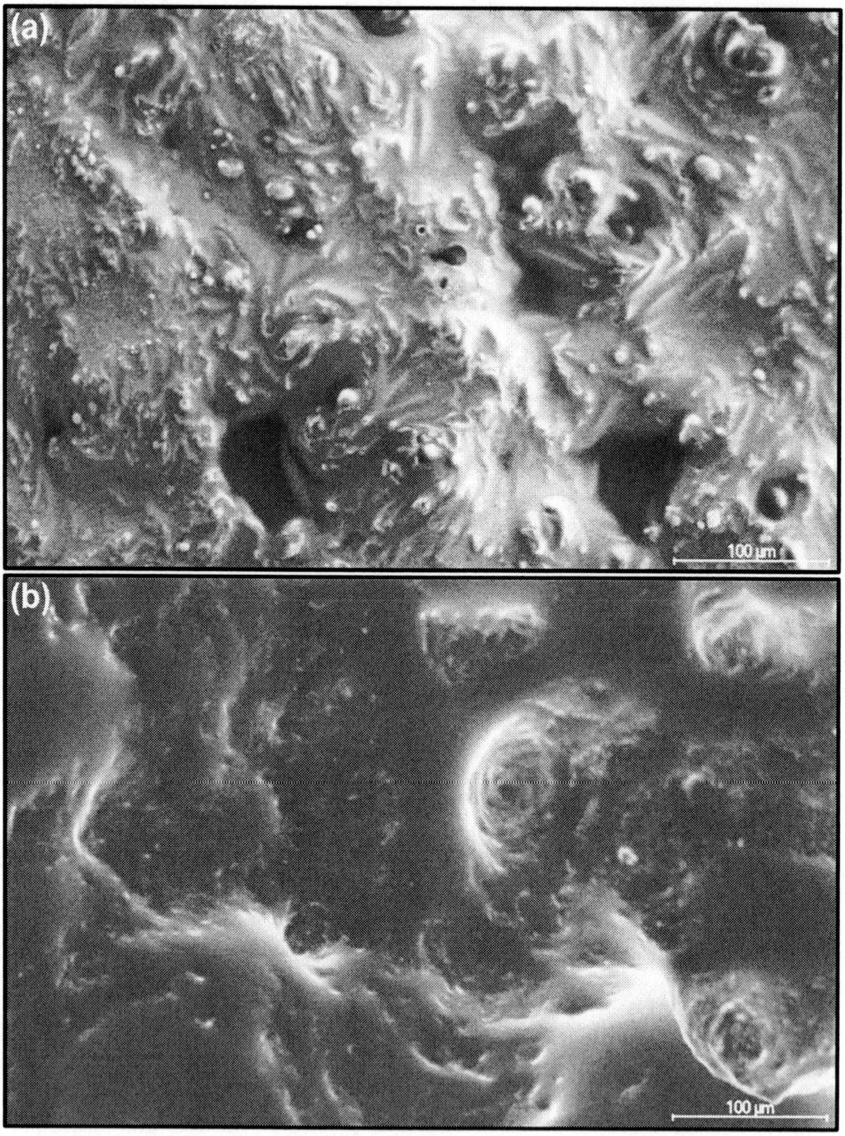

Figure 10. Microphotographs of mashed potatoes (a) Fresh sample without added cryoprotectants and with 50 g/kg added EVOO; (b) Fresh sample with added cryoprotectants and with 50 g/kg added EVOO; Magnification was 200 (bar = 100 μm).

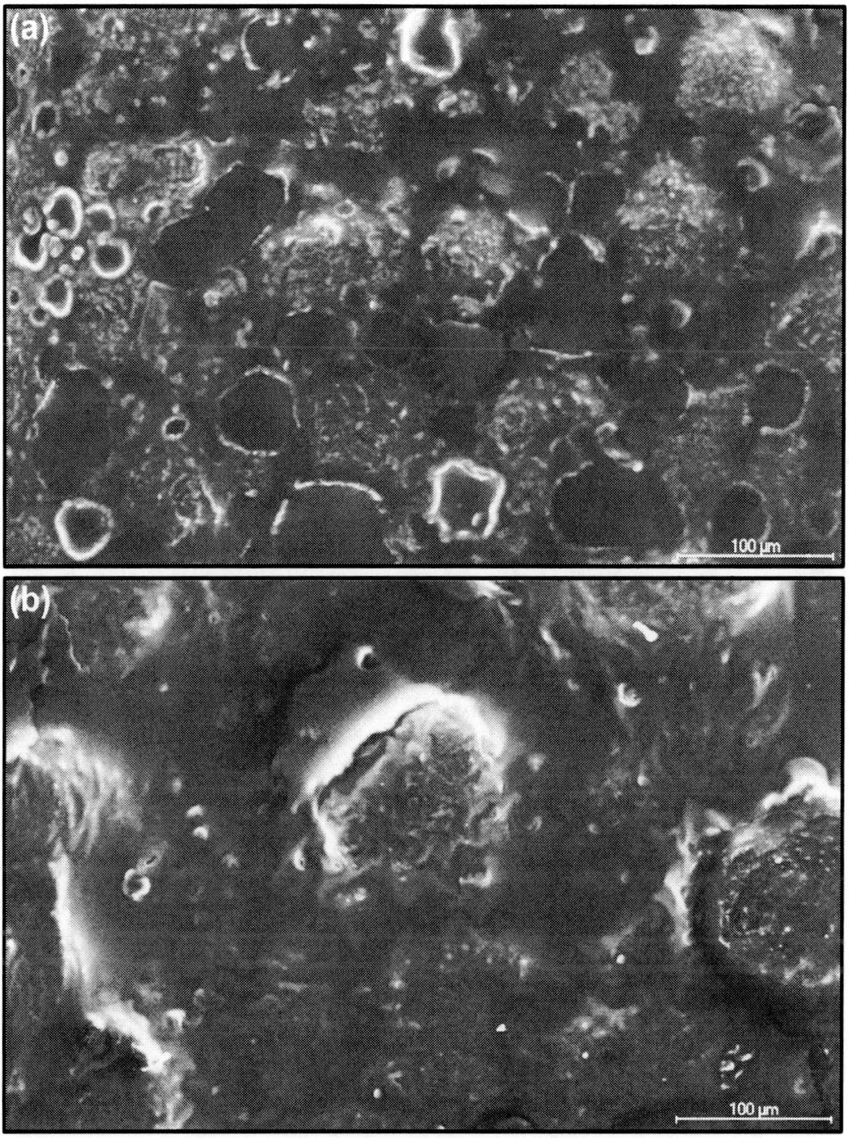

Figure 11. Microphotographs of mashed potatoes (a) Frozen/thawed sample without added cryoprotectants and with 50 g/kg added EVOO; (b) Frozen/thawed sample with added cryoprotectants and with 50 g/kg added EVOO; Magnification was 200 (bar = 100 µm).

Figure 10 shows microphotographs of the counterparts of the samples shown in the Figure 8, but with 50 g/kg added EVOO. When EVOO was added, a dispersed thin phase or layer of oil formed, enveloping all the microstructures constituting the MP. Figure 10a shows some oil droplets in the MPA sample, probably formed by aggregation through steric and/or electrostatic forces [6], whereas Figure 10b shows no oil droplets in presence of κ-C and XG. In MPA samples, freezing also had a negative influence on the formation of oil droplet clusters (Figure 11a; it is likely that the structural damage caused by freezing enabled the oil droplets to come close enough together to aggregate. Microphotograph of the processed sample with 50 g/kg added EVOO (Figure 11b) shows that white gel structures are also discernible in the presence of cryoprotectants.

Addition of XG to salad dressings induces depletion flocculation of the droplets and formation of a three dimensional weak gel network structure that retards the process of droplet creaming [30]. Adding a hydrocolloid causes protein-coated droplets to aggregate and be excluded from the region of continuous phase between them. Therefore, in the MP with added κ-C and XG and oil, the XG may have been adsorbed onto the surface of the droplets, enhancing stability against flocculation and coalescence and forming the white film observed in both microphotographs (Figures 10b, 11b). On the other hand, there are no noticeable differences between FMP and F/TMP samples with added κ-C and XG and oil, confirming that the addition of κ-C and XG significantly reduced quality differences between FMP and their F/TMP counterparts.

CONCLUSION

The addition of either EVOO or cryoprotectants and processing significantly affected the physical, structural and sensory characteristics of MP, although the effect of EVOO concentration depended on the presence of cryoprotectants and on freezing/thawing. Increased EVOO concentration resulted in less structured systems and enhancement of color due to an increase in overall light scattering and pigment content. Addition of κ-C and XG improved thickness, possibly through the exclusion effect of swollen starch granules promoting gelation of the κ-C. Addition of EVOO in increasing concentrations enhanced the sensory quality of MP in terms of reduced granularity, denseness, cohesiveness, adhesiveness and fibrousness, and increased homogeneity, ease of swallowing and palate coating. Instrumental texture measurements were able to distinguish the variations in mechanical textural attributes scored by the panellists. Conversely, geometrical textural attributes (granularity, homogeneity and fibrousness) have to be support by structural traits. Creaminess was the most crucial factor for OA of the products and could be explained by the presence of EVOO aggregates observed by microstructure analysis. Samples with 50 g/kg added EVOO were judged the best of all. There is a possibility of using EVOO in combination with MP to provide a highly nutritious product with improved physicochemical, functional and sensory characteristics.

Acknowledgments

The authors wish to thank the Spanish Ministry of Economy, Industry and Competitiveness for financial support (AGL2011-28569).

References

[1] Whybrow, S., Harrison, C. L. S., Mayer, C., & Stubbs, R. J. (2006). Effects of added fruits and vegetables on dietary intakes and body weight in Scottish adults. *British Journal of Nutrition*, 95, 496–503.

[2] Alvarez, M. D., Fernández, C., & Canet W. (2009). Enhancement of freezing stability in mashed potatoes by the incorporation of kappa-carrageenan and xanthan gum blends. *Journal of the Science of Food and Agriculture*, 89, 2115-2127.

[3] Boskou, D., & Visioli, F. (2003). Biophenols in olive oil and olives. In Bioavailability of micronutrients and minor dietary compounds. *Metabolic and technological aspects* (M. P. Vaquero, T. Garcia-Arias, A. Carbajal & F. J. Sanchez-Muniz, eds.) pp. 161–169, Research Signpost, Kerala.

[4] Mildner-Szkudlarz, S., & Jeleń, H. H. (2010). Detection of olive oil adulteration with rapeseed and sunflower oils mos electronic nose and SMPE-MS. *Journal of Food Quality*, 33, 21-41.

[5] Muniz, F. J. S. (2007). Olive oil, life key in the Mediterranean area. *Anales de la Real Academia Nacional de Farmacia*, 73(3), 653-692.

[6] Paraskevopoulou, A., Boskou, D., & Kiosseoglou, V. (2005). Stabilization of olive oil- lemon juice emulsion with polysaccharides. *Food Chemistry*, 90, 627-634.

[7] Mandala, I. G., Savvas, T. P., & Kostaropoulos, A. E. (2004). Xanthan and locust bean gum influence on the rheology and structure of a white model-sauce. *Journal of Food Engineering*, 64, 335–342.

[8] Dickinson, E., & Chen, J. S. (1999). Heat-set whey protein emulsion gels: Role of active and inactive filler particles. *Journal of Dispersion Science and Technology*, 20, 197–213.

[9] Chantrapornchai, W., Clydesdale, F., & Mcclements, D. J. (1999). Influence of droplet characteristics on the optical properties of colored oil-in-water emulsions. *Colloid Surface A*, 155, 373-382.

[10] Criado, M. N., Romero, M. P., Casanovas, M., & Motilva, M. J. (2008). Pigment profile and colour of monovarietal virgin olive oils from Arbequina cultivar obtained during two consecutive crop seasons. *Food Chemistry*, 110, 873-880.

[11] Ayuso, J., Haro, M. R., & Escolar, D. (2004). Simulation of the visible spectra for edible virgin oils: Potential uses. *Applied Spectroscopy*, 7, 474–480.

[12] Lamberti, M., Geiselmann, A., Conde-Petit, B., & Escher, F. (2004). Starch transformation and structure development in production and reconstitution of potato flakes. *LWT-Food Science and Technology*, 37, 417–427.

[13] Fernández, C., Canet, W., & Alvarez, M. D. (2009). Quality of mashed potatoes: effect of adding blends of kappa-carrageenan and xanthan gum. *European Food Research and Technology,* 229, 205-222.

[14] Severini, C., De Pilli, T., & Baiano, A. (2003). Partial substitution of pork backfat with extra-virgin olive oil in 'salami' products: effects on chemical, physical and sensorial quality. *Meat Science*, 64, 323-331.

[15] Alvarez, M. D., Fernández, C., & Canet, W. (2005). Effect of freezing/thawing conditions and long-term frozen storage on the quality of mashed potatoes. *Journal of the Science of Food and Agriculture*, 85, 2327-2340.

[16] Alvarez, M. D., Canet, W., & Fernández, C. 2008. Effect of addition of biopolymers on the mechanical properties, colour and sensory attributes of fresh and frozen/thawed mashed potatoes. *European Food Research and Technology*, 226, 1525-1544.

[17] Fernández, C., Alvarez, M. D., & Canet, W. (2008). Steady shear and yield stress of fresh and frozen/thawed mashed potatoes: effect of biopolymers addition. *Food Hydrocolloids*, 22, 1381-1395.

[18] Osorio, F. A., & Steffe, J. F. (1987). Back extrusion of power law fluids. *Journal of Texture Studies*, 18, 43-63.

[19] Kim, H. R., & Eliasson, A. C. (1993). Changes in rheological properties of hydroxypropyl potato starch pastes during freeze-thaw treatments II. Effect of molar substitution and cross-linking. *Journal of Texture Studies*, 24, 199-213.

[20] UNE 87025 (1996). *Manual de análisis sensorial.* Tomo I–Alimentación. Aenor, Madrid, pp. 167–186.

[21] International Organization for Standardization (ISO). (1993). International Standard 8586-1. *Sensory Analysis – General Guidance for the Selection, Training and Monitoring of Assessors – Part 1: Selected Assessors*, ISO, Geneva, Switzerland.

[22] International Organization for Standardization (ISO). (2003). International Standard 13299. *Sensory analysis – Methodology –*

General guidance for establishing a sensory profile, ISO, Geneva, Switzerland.

[23] Alvarez, M. D., Fernández, C., & Canet, W. (2011). Effect of cryoprotectant mixtures on rheological properties of fresh and froze/thawed mashed potatoes. *Journal of Food Process Engineering*, 34, 224–250.

[24] Mandala, I. G., & Palogou, E. D. (2003). Effect of preparation conditions and starch/xanthan concentration on gelation process of potato starch systems. *International Journal of Food Properties*, 6(2), 311–328.

[25] Eliasson, A. C., & Kim, H. R. (1992). Changes in rheological properties of hydroxypropyl potato starch pastes during freeze-thaw treatments I. A rheological approach for evaluation of freeze-thaw stability. *Journal of Texture Studies*, 23, 279-295.

[26] Sun, C., & Gunasekaran, S. (2009). Effects of protein concentration and oil-phase volume fraction on the stability and rheology of menhaden oil-in-water emulsions stabilized by whey protein isolate with xanthan gum. *Food Hydrocolloids*, 23, 165-174.

[27] Cavella, S., Di Monaco, R., Torrieri, E., & Masi, P. (2009). Structure of a new functional walnut oil-enriched mayonnaise. 9th International Conference on Chemical and Progress Engineering, pts 1-3. *Chemical Engineering Transactions*, 17, 879-884.

[28] Giannouli, P., & Morris, E. R. (2003). Cryogelation of xanthan. *Food Hydrocolloids*, 17, 495-501.

[29] Arocas, A., Sanz, T., & Fiszman, S. M. (2009). Improving effect of xanthan and locust bean gums on the freeze-thaw stability of white sauces made with different native starches. *Food Hydrocolloids*, 23, 2478-2484.

[30] Parker, A., Gunning, P.A., NG, K., & Robins, M. M. (1995). How does xanthan stabilise salad dressing? Food *Hydrocolloids*, 9, 333–342.

[31] Baranowska, H. M., Sikora, M., Kowalski, S., & Tomasik, P. (2008). Interactions of potato starch with selected polysaccharide

hydrocolloids as measured by low-field NMR. *Food Hydrocolloids*, 22, 336-345.

[32] Ferrero, C., Martino, M. N., & Zaritzky, N. E. (1994). Corn starch–xanthan gum interaction and its effect on the stability during storage of frozen gelatinized suspensions. *Starch-Stärke*, 46(8), 300–308.

[33] Bemiller, J. N., & Whistler, R. L. (1996). Carbohydrates. In *Food chemistry* (O. R. Fennema, ed.) pp. 157–224, Marcel Dekker, New York.

[34] Liehr, M., & Kuliche, W. M. (1996). Rheological examination of the influence of hydrocolloids on the freeze-thaw-stability of starch gels. *Starch-Stärke*, 48, 52-57.

[35] Christianson, D. D., Hodge, J. E., Osborne, D., & Detroy, R. W. (1981). Gelatinization of wheat starch as modified by xanthan gum, guar gum, and cellulose gum. *Cereal Chemistry*, 58(6), 513–517.

[36] De Wijk, R. A., Van Gemert, L. J., Terpstra, M. E. J., & Wilkison, C. L. (2003). Texture of semi-solids; sensory and instrumental measurements on vanilla custard desserts. *Food Quality and Preference*, 14, 305-317.

[37] Szczesniak, A. S. (2002). Texture is a sensory property. *Food Quality and Preference*, 13, 215–225.

[38] Juszack, L., Fortuna, T., & Kosla, A. (2003). Sensory and rheological properties of Polish commercial mayonnaises. *Nahrung*, 47, 232-235.

In: Olive Oil ISBN: 978-1-53612-563-4
Editors: T. Fritjof and B. Henning

Chapter 5

Mediterranean Virgin Olive Oils: The Relationship between Volatile Compounds and Sensory Characteristics

Imen Oueslati* and Hédia Manai
Laboratoire de Biotechnologie de l'olivier,
Centre de Biotechnologie de Borj-Cédria, Hammam-Lif, Tunisia

Abstract

Virgin olive oils (VOOs) represent the main source of fats in the countries of the Mediterranean basin where the production of olive oil is concentrated. Its quality is based upon consumer perceptions of aroma, and the absence of sensory defects. Volatile compounds responsible for the olive oil aroma count approximately two hundred compounds belonging to several chemical classes, identified via gas chromatography-mass spectrometry. The main characteristics of the volatile compounds of the several Mediterranean olive oil categories and the several enzymes involved in the pathways and in their activity were investigated.

* Corresponding Author Email: imen.oueslati@fst.rnu.tn; amounadream@yahoo.fr.

The synergy and antagonism processes between volatile compounds contribute to the sensory evaluation of VOOs, so it is of great interest to ascertain the relationships between sensory attributes and the volatiles responsible for them. Aroma compounds responsible for flavours, and for off-flavours including factors affecting the volatile fraction, are carefully examined, also considering the causes that give rise to off-flavours. New approaches were adopted to the definition of EVOO quality that might include also healthy and sensory parameters and on the agronomic and technological factors that most influence their variability.

Keywords: virgin olive oils, volatiles, sensory attributes, quality

INTRODUCTION

In the Mediterranean area, healthy, nutritional, and sensorial properties of olive oil have been known for many centuries. Olive oil is considered the main fat source of the Mediterranean diet, and it is appreciated for its characteristics such as: aroma, taste, color, and nutritive properties that are distinguishable from other vegetable oils.

Several studies (Nsir, et al., 2017; López de las Hazas, et al., 2017) have evidenced the beneficial effects on health of the Mediterranean virgin olive oil (VOO). Numerous research have proved that some non-volatile compounds, such as phenolic compounds, are related in an indirect way to positive effects on human health (López de las Hazas, et al., 2017), whereas sensory attributes are directly ascribable to the strong stimulation of human sensory receptors by volatile and some non-volatile compounds (Angerosa, et al., 1996) present in VOO. Phenolic compounds stimulate the tasting receptors and the free endings of the trigeminal nerve eliciting the former with the bitterness perception and the latter with pungency and astringency (Angerosa, et al., 2004). The combined effect of odour (directly via the nose or indirectly through a retronasal path, via the mouth), taste and chemical responses (as pungency) gives rise to the sensation generally perceived as "flavour." All compounds, strongly affecting sensory receptors, can decisively influence food acceptability, direct the preference of consumers and, in short, to a great extent, determine the quality of life.

Aroma compounds stimulate much more qualities and therefore are mainly responsible for the characteristic flavour of foods (Kesen, et al., 2014). Olive oil possesses a highly distinctive taste and flavour due to specific volatile organic compounds, belonging to several chemical classes, namely aliphatic and aromatic hydrocarbons, aliphatic and triterpenic alcohols, aldehydes, ketones, ethers, esters, furan and thiophene derivatives (Romero, et al., 2015; Kalua, et al., 2007). These compounds, retained by olive oil during the extraction process, stimulate human gustative and olfactive receptors giving rise to olive oil balanced flavour of green and fruity attributes. Such compounds stimulate the free endings of the terminal nerve located in all the palate and in the gustative buds promoting the chemesthetic perceptions of pungency, astringency and metallic attributes.

Olive oil aroma is mainly produced by oxidation of fatty acids. The endogenous plant enzymes, through the lipoxygenase (LOX) pathway, are responsible for the positive aroma perceptions in olive oil, whereas chemical oxidation and exogenous enzymes, usually from microbial activity, are associated with sensory defects. Both the processing and storage of the fruit and the oil contribute greatly to the flavour and overall quality of olive oil (Angerosa, 2002).

Antagonism and/or synergism among different volatile molecules can occur, affecting the global flavour of VOO, so it is of great interest to establish, in the present work, the relationships between sensory attributes and volatile compounds, and their influence on the VOO quality.

Virgin Olive Oil Quality and Volatile Compounds

There are several ways of defining *quality* and perhaps there is no single universal definition that adequately satisfies all situations. International Standards Organization (ISO 8402, 1986) defines *quality* as: "the totality of features and characteristics of a product or service that bears its ability to satisfy stated or implied needs". Gould (1992) defines the general terms of *quality* as: "the combination of attributes or characteristics of a product that have significance in determining the degree of acceptability of that product by the user". Olive oil quality may be defined from commercial, nutritional or organoleptic perspectives (Duran, 1990).

The large increase in demand for high-quality olive oil is thought to be related to its peculiar organoleptic characteristics that play an important role in human nutrition. In fact, all compounds, strongly affecting sensory receptors, can decisively influence food acceptability, direct the preference of consumers and, in short, to a great extent, determine the quality of life (De Santis, et al., 2015). For this reason, quality olive oil assessment involves evaluation of sensory characteristics, which was carried out according to the "panel test" method, a standardised sensory method that classifies oils on the basis of the presence or absence of standardised defects and of the presence of positive characteristics named "fruity," "bitter" and "pungent" (Procida, et al., 2016). Positive attributes and sensory defects in olive oil could be associated with the volatile and phenol profiles, which are responsible for aroma and taste, respectively. The qualitative and quantitative composition of the volatile compounds of high quality olive oils depends closely on the levels of enzymes involved in the pathways and on their activity (Procida, et al., 2016). Genetic characteristics fix the contents of the different enzymes and are therefore responsible for the qualitative composition of volatile compounds (Salas, 2004). The VOO with high quality contains mainly volatiles derived from linoleic and linolenic acid decomposition through the lipoxygenase pathway (Ridolfi, et al., 2002). The quantity of the different volatile compounds is related to the enzyme activities which, in oils of good quality, are dependent to the ripening degree and the sanitary conditions of fruits, the technological aspects, the temperature reached by the olive paste during crushing, the time and the temperature of malaxation and the kind of system used for oil extraction (Angerosa, 2002).

Moreover, the category of EVOO should not show any defects (e.g., fusty, musty, winey, metallic, rancid). A greater number of volatile compounds have been found in virgin olive oils of poorer quality. Some of these volatiles give rise to sensory defects when they are present at high levels (Kalua, et al., 2007). They are produced by over-ripening of the fruits. The uses of fruits that have partially degraded tissues cause an increase in enzymatic and microrganism activities and oxidative reactions; therefore, the produced oil probably will be characterised by a higher free acidity and perceivable sensory defects. When olives are accumulated in piles for many days, the high temperature and humidity inside the mass promotes proliferation of bacteria, yeasts, and moulds, producing undesirable fermentation and degradation that give rise to specific volatile molecules responsible for unpleasant odours (Angerosa, 2002).

Composition and Biosynthesis of the Volatile Fraction of Virgin Olive Oil

Olive oil aroma consists of a complex mixture of volatile compounds accounting more than two hundred volatiles in VOOs. They are hydrocarbons, alcohols, aldehydes, ketones, acids, esters, ethers, furan derivatives, thiophene derivatives, pyranones, thiols, and pyrazines. From this large number of compounds, only 67 were found to be present at levels higher than their odour threshold contribute to the aroma of VOO (Table 1) (Kalua, et al., 2007). However, it is well-accepted that only a small fraction on the large number of volatile compounds occurring in food actually contribute to the overall aroma (Kesen, et al., 2014).

Enzymatic reactions and autoxidation play crucial roles in the occurrence of aroma substances. The formation of these compounds starts at the moment of cell disruption during crushing of olives and continues during the extraction process.

Synthesis of the aroma compounds of VOO occurs through the LOX pathway (comprising mainly the actuation of LOX and hydroperoxide-lyase enzymes), and fatty acid or amino acid metabolism, as depicted in Figure 1 (Angerosa, et al., 2004; Angerosa, et al., 2002).

Table 1. Volatile compounds and sensory descriptors in olive oil (Kalua, et al., 2007; Morales et al., 2005; Aparicio and Luna, 2002; Aparicio and Morales, 1998; Reiners and Grosch, 1998)

Volatile compounds	Sensory descriptor (aroma)
Aldehydes	
Acetaldehyde	Pungent, sweet
3-methylbutanal	Malty
2-methylbutanal	Malty
Pentanal	Woody, bitter, oily
Trans-2-pentenal	Green, apple, bitter, almond
Hexanal	Green-sweet
Cis-3-Hexenal	Green, Leaf-like
Trans-2-Hexenal	Green, apple-like, Bitter almonds, astringent
Heptanal	Oily, fatty, woody
Trans-2-Heptenal	Oxidised, tallowy, pungent
2,4-heptadienal	Fatty, rancid
Octanal	Fatty, sharp, Citrus-like
Trans-2-Octenal	Herbaceous, spicy
Nonanal	Fatty, waxy, pungent
Trans,trans-2,4-Nonadienal	Soapy, penetrating, Deep-fried
Cis-2-Nonenal	Green, fatty
Trans-2-Nonena	Paper-like, fatty
Decanal	Penetrating, sweet, waxy
Trans-2-Decenal	Painty, fishy, fatty
2,4-decadienal	Strong, fatty
Trans,trans-2,4-Decadienal	Deep-fried
Trans,cis-2,4-Decadienal	Deep-fried
Trans-4,5-Epoxy-trans-2- decanal	Metallic
Alcohols	
Ethanol	Alcohol
Butan-2-ol	Winey
2-Methyl butan- 1 -ol	Winey, spicy
3-Methyl butan- 1-ol	Woody, whiskey, sweet
Pentanol	Fruity, Strong, sticky, balsamic
3-Penten-2-ol	perfumery, woody
Hexanol	Fruit, banana, soft, Undesirable
Trans-2-Hexen-1-ol	Green grass, leaves, sweet
Trans-3-Hexen-1-ol	Green
Cis-3-Hexenol	Green, Leaf-like
Heptan-2-ol	Earthy, sweefy
6-Methyl-5-hepten-3-ol	Perfumey, nutty
Esters	
Ethyl acetate	Sticky, sweet
Butyl acetate	Green, fruity, pungent
Hexyl acetate	Green, fruity, sweet
Cis-3-Hexenyl acetate	Green, Banana-like
Ethyl propanoate	Fruit, strong
Ethyl butanoate	Sweet, fruity, Cheesy
Ethyl isobutyrate	Fruity
Propyl butanoate	Pineapple, sharp

Table 1 (Continued)

Volatile compounds	Sensory descriptor (aroma)
2-Methylpropyl butanoate	Unpleasant, winey, Fusty
Ethyl 2-methylbutyrate	Fruity
Ethyl 3-methylbutyrate	Fruity
Ethyl cyclohexylcarboxylate	Aromatic, fruity
Ketones	
Butan-2-one	Ethereal, fruity
l-Penten-3-one	Green, pungent
Heptan-2-one	Sweet, fruity
6-Methyl-5-hepten-2-one	Pungent, green
Octan-2-one	Mould, green
l-Octen-3-one	Mushroom, mould, pungent;
cis-1,5-Octadien-3-one	Geranium-like
trans-b-Damascenone	Boiled apple-like
Carboxylic acids	
Acetic acid	Sour, vinegary
Propanoic acid	Pungent, sour
Butanoic acid	Rancid, cheese
3-Methylbutyric acid	Sweaty
Pentanoic acid	Inpleasant, pungent
Hexanoic acid	Pungent, rancid, fatty
Octanoic acid	Oily, fatty
Other compounds	
Octane	Sweety, alcane
4-Methoxy-2-methyl-2- butanethiol	Black currant-like, catty
Guaiacol	Phenolic, burnt

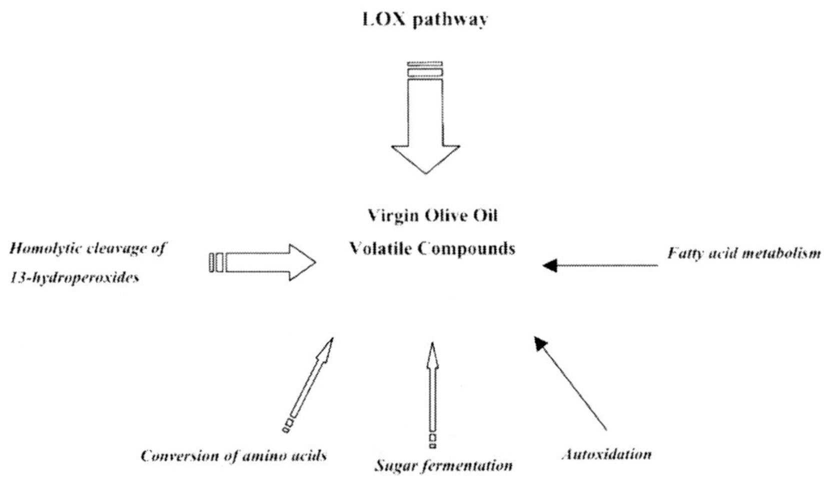

Figure 1. The enzymatic and chemical paths involved in the production of virgin olive oil volatile compounds (Angerosa, et al., 2004).

The most abundant aroma compounds in VOO are produced enzymatically from polyunsaturated fatty acids through the LOX pathway. It is stated that in the aroma of these oils C_6 aldehydes, C_6 alcohols and their corresponding esters are the most abundant accumulation products. Moreover, reasonable amounts of C_5 carbonyl compounds, C_5 alcohols and pentene dimers were identified in the aromatic profile of VOO (Angerosa, et al., 1998).

The number of volatile compounds detected in the aroma of an olive oil depends on the methodology adopted for their determination and on the quality of the VOO (Cavalli, et al., 2003). The formation of the C_6 and C_5 compounds is affected by the cultivar, the degree of fruit ripeness, the storage time of fruits prior to oil extraction, and by the processing (Baccouri, et al., 2008; Angerosa, et al., 2000; Aparicio, and Luna, 2002; Morales, et al., 2005).

The best way to comprehend aroma development is the understanding of the various enzyme pathways, which provide the formation of olive oil volatile compounds. In general, the release of enzymes at the cell destruction stage can be considered the switch-on of the LOX pathway. The reaction pathway is initiated by the hydrolysis of triglycerides and phospholipids, mediated by acyl hydrolase (AH), leading to the release of fatty acids. LOX, after their release, become immediately active and transform the unsaturated fatty acids, produced by the action of AH, linolenic (LnA) and linoleic (LA) acids, into their corresponding 9- and 13-hydroperoxides, as shown in Figure 2. The enzyme is more active with linolenic acid than linoleic acid by a factor of two (Salas, et al., 1999; Sanchez, and Salas, 2000). The higher LOX activity for linolenic acid than linoleic acid supports the biogenesis of more of the six-carbon unsaturated volatile compounds (Figure 2), which are the major constituents of the aroma VOO with high quality (Salas, et al., 1999).

The subsequent cleavage of fatty acids 13-hydroperoxides is catalysed by specific hydroperoxide lyases (HPL) leading to the formation of C_6 aldehydes ((*Z*)-hex-3-enal and hexanal from linolenic and linoleic acids, respectively) and oxoacids. The unsaturated form of C_6 aldehyde ((*Z*)-hex-3-enal) undergo rapid isomerisation to the more stable (*E*)-hex-2-enal.

The action of alcohol dehydrogenase (ADH), catalyses the reversible reduction of aliphatic C_6 aldehydes to the corresponding volatile. Alcohol species are further transformed into esters due to the catalytic activity of alcohol acetyl transferase (AAT), producing acetates (Kalua, et al., 2007) (Figure 2). When the substrate is LnA, LOX catalyses, besides the hydroperoxide formation, also its cleavage, via an alkoxyradical, increasing the formation of stabilised pent-1,3-diene radicals. These compounds can suffer dimerisation leading to the production of C_{10} hydrocarbons (pentene dimmers) or react with a hydroxyl radical present in the medium, leading to C_5 carbonyl compounds (Pizarro, et al., 2011).

The presence of other volatile compounds, namely C_7-C_{11} monounsaturated aldehydes, C_6-C_{10} dienals, C_5 branched aldehydes and alcohols and some C_8 ketones, in relatively high concentrations, in the aroma of VOO, is associated with unpleasant notes.

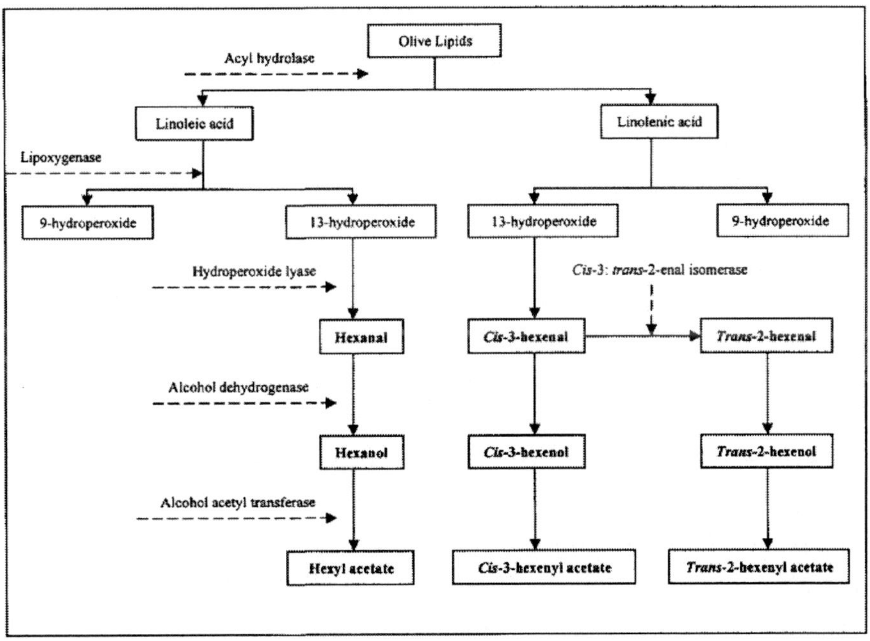

Figure 2. Lipoxygenase pathways involved in the production of C_6 and C_5 volatile compounds (Olias, et al., 1993; Ridolfi, et al., 2002).

The action of amino acids in olive oil volatile production has been paid little attention. It is established that valine and leucine are converted to volatile compounds, including methyl-branched alkyl and acyl compounds of esters, and into methyl-branched alcohols, which have the potential to change the sensory perception (Tressl, and Drawert, 1973). The presence of other minor volatile compounds may provide useful quality markers and lead to an improved understanding of the formation or degradation of the major volatile compounds (Buttery, and Takeoka, 2004). Moreover, an understanding of the various pathways can improve the production and storage of premium quality olive oil.

Flavours and Off-Flavours of Virgin Olive Oil

The volatile compounds formed during the processing of olive fruit contribute to a combined sensation of smell and taste, commonly called flavour. Evaluation of the sensory quality of VOOs involves perception of both favourable and unfavourable sensory attributes, with evaluation of sensory defects being used to classify oils into various grades (Kalua, et al., 2007). More than 120 volatile compounds that contribute both positively and negatively to the sensory properties of olive oil have been identified (Aparicio & Luna, 2002). Some of the common volatile compounds, with their sensory characterisation and odour threshold, are arranged, based on their functional groups and carbon number. The major volatile compounds of olive oil which contribute for the positive attributes of olive oil aroma (fruity, pungent and bitter) include C_6 aldehydes (hexanal, (E)- hex-2-enal), alcohols (hexan-1-ol and 3- methylbutan-1-ol), and their corresponding esters and minor amounts of C_5 carbonyl compounds, alcohols and pentene dimers are responsible for pleasant notes. Their concentrations, except for (E)-hex-2-enal, varying widely, are generally very low reaching minimum levels of ppb. Thus, volatile compounds, which are responsible for most sensory properties of olive oils, play a significant role on the evaluation of the overall oil quality having a decisive influence on acceptability (Boskou, 2006).

The off-flavour compounds associated with unpleasant sensory notes can be assembled in five classes- fusty, moistness- humidity, winey-vinegary, metallic and rancid (Morales, et al., 2005; Escuderos, et al., 2007; Angerosa, 2002; Kalua, et al., 2007). The main defects or off-flavours are due to sugar fermentation (winey), amino acid (leucine, isoleucine, and valine) conversion (fusty), enzymatic activities of moulds (musty) or anaerobic microorganisms (muddy), and to auto-oxidative processes (rancid) (Bendini, et al., 2012). Chemical oxidation is responsible for the formation of off-flavour compounds, such as pent-2-enal and hept-2-enal. Moistness-humidity, which possesses the highest sensory significance, is related to the presence of C_8 volatile compounds (e.g., oct-1-en-3-ol and to a lesser extent oct-1-en-3-one) and short chain fatty acids (Morales, et al., 2005). Pentanal, hexanal, octanal and nonanal are the major compounds (Kiritsakis, 1998; Morales, and Aparicio, 1997) and carboxylic acid, such as hexanoic, and propanoic acid, have also been detected during the oxidation (Vichi, et al., 2003). Other volatile compounds detected in the late oxidation stages are 2-pentylfuran and 2-ethylfuran, which might be useful in distinguishing oxidation at late stages (Vichi, et al., 2003). All these compounds can be considered as markers of oxidation except for hexanal (Vichi, et al., 2003). The amount of hexanal does not distinguish oxidised oils from virgin oils as this compound originates from both the enzymatic and chemical oxidation pathways (Morales, and Aparicio, 1997; Vichi, et al., 2003). The rate of volatile compound formation during oxidation has been considered in choosing oxidation markers.

In addition, antagonism and/or synergism among different molecules can occur, affecting the global flavour of EVOO. Chemical factors of molecules (volatility, hydrophobic character, size, shape, conformational structure), type and position of functional groups as well as external factors (such as matrix effects), appear to affect the odour and taste intensity more than their concentration due to their importance in establishing bonds with receptor proteins (Angerosa, et al., 2004).

Beside the volatile compounds, the sensory attributes of EVOO depend also on the content of some other minor components, such as phenolic compounds. Polyphenols contribute to the organoleptic properties of VOOs and are commonly described as bitter and astringent. Less commonly, polyphenols are associated with pungency, that is, peppery, burning, or hot sensations (Tsimidou, 1998). Few individuals, except for trained tasters of EVOO, know that the bitterness and pungency perceived are considered positive attributes. Several investigations (Gutiérrez-Rosales, et al., 2003) have demonstrated that some phenols, and in particular secoiridoid derivatives of hydroxytyrosol, are the main contributors to the bitterness of olive oil; other phenolic molecules such as decarboxymethyl-ligstroside aglycone, which seems to be a key source of the burning sensation, can stimulate the free endings of the trigeminal nerve located in the palate and gustative buds giving rise to the chemesthetic perceptions of pungency and astringency (Andrewes, et al., 2003).

Consumer Acceptability of the Sensory Properties of Virgin Olive Oil

The beliefs and attitudes of individuals with regards to food products are strongly dependent on their cultural traditions and on their education and culinary habits and they may vary with information (Caporale, et al., 2006). According to some research (Fishbein, 1967; Lancaster, 1966), consumers always relate a product with a set of attributes. Therefore, consumers take into account multiple features of the products and use them to build into their minds a model of preferences. Based on the selected model, the consumer finally decides which product to purchase and which products to reject.

Extra virgin olive oil is a typical Mediterranean production whose typicality is strongly affected by the origin of its raw material and the manufacturing technology (Caporale, et al., 2006). Generally, the Mediterranean consumers (i.e., Spanish, Italian, Greek, and Tunisian consumers) appreciate what is familiar, what is strongly linked to the territory (tradition/origin) (Caporale, et al., 2006; Costell, et al., 2010), so, the quality of VOO is not always correctly perceived. Caporale, et al. (2006), Fotopoulos, and Krystallis (2001), and Stefani, et al. (2006) showed that consumers who are experienced, local, or familiar with a particular region of origin tend to consider region a key factor that drives purchasing and preference, while these factors do not seem to influence urban, less knowledgeable, and less experienced consumers. The new olive oil consumers (i.e., United State, Russian, Chinese, and Australian consumers) are interested in olive oil for two main reasons: health benefits and flavour (Santosa, 2010).

For a correct perception of the overall quality, the fruity (green or ripe), bitter and pungent attributes should be perceived by consumers as "healthy" indicators of quality (Carluccio, et al., 2003). However, consumers do not appear to practice an informed consumption of EVOO. A study in Italy (Caporale, et al., 2006) demonstrated that information evoking the origin of EVOO create a favourable hedonic expectation, regarding specific sensory attributes, such as pungency and bitterness. Interestingly, information on origin did not have the same effect on different consumer responses such as liking or typicality. Although, it was shown that information about origin affects the expectations with regards to specific sensory attributes in familiar consumers. A consumer study was conducted in US, to evaluate preferences and attitudes regarding EVOO, showed that, for the majority of consumers, bitterness and pungency were negative drivers of liking. Properties that drove positive ratings were fruity (green and ripe), nutty, and tea-like flavours. The EVOO's price, available information, and reputation were key factors that drove purchases in this consumer population (Delgado, and Guinard, 2011).

CONCLUSION

In accordance with what has been stated above, we can conclude that from healthy olives harvested at the optimum ripening stage and properly processed, it is always possible to obtain VOO with high quality (independent of the genetic factor), characterised by the abundance of the LOX products (especially C_6 aroma products) which contribute to the positive attributes of olive oil aroma (fruity, pungent and bitter). However, from unhealthy and damaged olives, or from those harvested at an over ripe stage, it is inevitable to produce an olive oil characterised by an off-flavour and sensory defects. Thus, both natural (olive variety, environmental conditions, degree of ripening and health status of olives) and extrinsic (technological processing by olive farmer/mill worker) factors may profoundly influence olfactory and gustative notes.

REFERENCES

Andrewes, P., Busch, J. L. H. C., De Joode, T., Groenewegen, A. and Alexandre, H. (2003). Sensory properties of virgin olive oil polyphenols: identification of deacetoxy ligstroside aglycon as a key contributor to pungency. *Journal of Agriculture and Food Chemistry*, 51: 1415-1420.

Angerosa, F. and Di Giovacchino, L. (1996) Natural Antioxidants of Virgin Olive Oil Obtained by Two and Tri-Phase Centrifugal Decanters. *Grasas y Aceites,* 47: 247-254.

Angerosa, F., d'Alessandro, N., Basti, C. and Vito R. (1998). Biogeneration of volatile compounds in virgin olive oil: their evolution in relation to malaxation time. *Journal of Agricultural and Food Chemistry*, 46: 2940- 2944.

Angerosa, F. (2002). Influence of volatile compounds on virgin olive oil quality evaluated by analytical approaches and sensor panels. *European Journal of Lipid Science and Technology*, 104(9–10): 639-660.

Angerosa, F., Servili, M., Selvaggini, R., Taticchi, A., Esposto, S. and Montedoro, G. F. (2004). Volatile Compounds in Virgin Olive Oil: Occurrence and Their Relationship with the Quality. *Journal of Chromatography A*, 1054: 17-31.

Angerosa, F., Mostallino, R., Basti, C. and Vito, R. (2000). Virgin Olive Oil Odour Notes: Their Relationships with Volatile Compounds from the Lipoxygenase Pathway and Secoiridoid Compounds. *Food Chemistry*, 68: 283-287.

Andrewes, P., Busch, J. L. H., De Joode, T., Groenewegen, A. and Alexandre, H. (2003). Sensory Properties of Virgin Olive Oil Polyphenols: Identification of Deacetoxy-Ligstrosideaglycon as a Key Contributor to Pungency. *Journal of Agricultural and Food Chemistry*, 51:1415-1420.

Aparicio, R. and Luna, G. (2002). Characterisation of monovarietal virgin olive oils. *European Journal of Lipid Science and Technology*, 104(9–10): 614–627.

Baccouri, O., Bendini, A., Cerretani, L., Guerfel, M., Baccouri, B., Lercker, G., Zarrouk, M. and Daoud, D. (2008). Comparative study on volatile compounds from Tunisian and Sicilian monovarietal virgin olive oils. *Food Chemistry,* 111: 322–328.

Bendini, A., Valli, E., Barbieri, S., and Toschi, T. G. (2012). Sensory analysis of virgin olive oil, in *Olive Oil – Constituents, Quality, Health Properties and Bioconversions* (ed. B. Dimitrios), InTech, Rijeka, Croatia, pp. 109–130.

Boskou, D. (2006). Olive oil composition. In: *Olive Oil: Chemistry and Technology*. Ed. D. Boskau, 2nd edition AOCS Press, Champaign, IL, USA, pp. 41–7.

Buttery, R. G. and Takeoka, G. R. (2004). Some unusual minor volatile components of tomato. *Journal of Agricultural and Food Chemistry*, 52(20): 6264–6266.

Caporale, G., Policastro, S., Carlucci, A. and Monteleone, E. (2006). Consumer Expectations for Sensory Properties in Virgin Olive Oils. *Food Quality and Preference*, 17: 116-125.

Carluccio, M. A., Siculella, L., Ancora, M. A., Massaro, M., Scoditti, E., Storelli, C., Visioli, F., Distante, A. and De Caterina, R. (2003). Olive Oil and Red Wine Antioxidant Polyphenols Inhibit Endothelial Activation: Antiatherogenic Properties of Mediterranean Diet

Phytochemicals. *Arteriosclerosis, Thrombosis, and Vascular Biology,* 23: 622-629.

Cavalli, J. F., Fernandez, X., Lizzani-Cuvelier, L., and Loiseau, M. (2003). Comparison of static headspace, headspace solid phase microextraction, headspace sorptive extraction, and direct thermal desorption techniques on chemical composition of French olive oils. *Journal of Agricultural and Food Chemistry*, 51: 7709–7716.

Costell, E., Tárrega, A. and Bayarri, S. (2010). Food Acceptance: The Role of Consumer Perception and Attitudes. *Chemosensory Perception*, 3: 42-50.

Delgado, C. and Guinard, J. X. (2011). How do consumer hedonic ratings for extra virgin olive oil relate to quality ratings by experts and descriptive analysis ratings?. *Food Quality and Preference*, 22: 213–225.

De Santis, D., and Frangipane, M. T. (2015). Sensory Perceptions of Virgin Olive Oil: New Panel Evaluation Method and the Chemical Compounds Responsible. *Natural Science*, 3: 132-142.

Duran, R. M. (1990). Relationship between the composition and ripening of the olive and the quality of the oil. *Acta Horticulturae*, 286: 441–451.

Escuderos, M. E., Uceda, M., Sánchez, S. and Jiménez, A. (2007). Instrumental technique evolution for olive oil sensory analysis. *European Journal of Lipid Science and Technology,* 109: 536–546.

Fishbein, M. (1967). "Attitudes and prediction of behavior," in Fishbein, M. (Ed.), *Readings in Attitude Theory and Measurement*, Wiley, New York, NY, pp. 477-92.

Fotopoulos, C. and Krystallis, A. (2001). Are quality labels a real marketing advantage? A conjoint application on Greek PDO protected olive oil. *Journal of International Food and Agribusiness Marketing*, 12(1) : 1–22.

Gould, W. A. (1992). *Total quality management for the food industries.* Baltimore, USA: CTI Publications Inc.

Gutiérrez-Rosales, F., Rios, J. J. and Gomez-Rey, MA. L. (2003). Main polyphenols in the bitter taste of virgin olive oil. Structural confirmation by on-line High-Performance Liquid Chromatography Electrospray Ionization Mass Spectrometry. *Journal of Agriculture and Food Chemistry*, 51: 6021-6025.

Kalua, C. M., Allen, M. S., Bedgood, D. R., Bishop, A. G., Prenzler, P. D. and Robards, K. (2007). Olive Oil Volatile Compounds, Flavour Development and Quality: A Critical Review. *Food Chemistry*, 100: 273-286.

Kesen, S., Kelebek, H. and Selli, S. (2014). Characterization of the Key Aroma Compounds in Turkish Olive Oils from Different Geographic Origins by Application of Aroma Extract Dilution Analysis (AEDA). *Journal of Agriculture and Food Chemistry,* 62: 391–401.

Kiritsakis, A. K. (1998). Flavour components of olive oil – a review. *Journal of the American Oil Chemists' Society*, 75: 673–681.

Lancaster, K. A. (1966). "New approach to consumer theory," *Journal of Political Economy,* Vol. 74 No. 2, pp. 132-57.

López de las Hazas, M. C., Piñol, C., Macià, A. and Motilva, M. J. (2017). Hydroxytyrosol and the Colonic Metabolites Derived from Virgin Olive Oil Intake Induce Cell Cycle Arrest and Apoptosis in Colon Cancer Cells. *Journal of Agriculture and Food Chemistry, In press.*

Morales, M. T., and Aparicio, R. (1997). Changes in the volatile composition olive oil during oxidation: flavours and off-flavours. *Journal of Agriculture and Food Chemistry*, 45: 2666–2673.

Morales M., Luna, G. and Aparicio R. (2005). Comparative Study of Virgin Olive Oil Sensory Defects. *Food Chemistry,* 91: 293-301.

Nsir, H., zychlinska, M. N., Cardile, V., Eleonora Graziano, A. C., Avola, R., Esafi, H, Bendini, A., Zarrouk, M., Loreto, C., Rapisarda, V., Castrogiovanni, P. and Musumeci, G. (2017). Polar and apolar extra virgin olive oil and leaf extracts as a promising antiinflammatory natural treatment for osteoarthritis. *Acta Histochemica*, 119: 407-416.

Olias, J. M., Perez, A. G., Rios, J. J. and Sanz, L. C. (1993). Aroma of virgin olive oils biogenesis of the green odor notes. Journal of Agricultural and Food Chemistry. 41, 2368–2373. on extra virgin olive oil quality. *Journal of the American Oil Chemists' Society*, 86: 1137-1140.

Pizarro, C., Tecedor, S. R., Pérez-del-Notario, N. and Sáiz, J. M. G. (2011). Recognition of Volatile Compounds as Markers in Geographical Discrimination of Spanish Extra Virgin Olive Oils by Chemometric Analysis of Non-Specific Chromatography Volatile Profiles. *Journal of Chromatography A,* 1218: 518–523.

Procida, G., Cichelli, A., Lagazioc, C. and Conte, L. S. (2016). Relationships between volatile compounds and sensory characteristics in virgin olive oil by analytical and chemometric approaches. *Journal of the Science of Food and Agriculture*, 96: 311–318.

Reiners, J., and Grosch, W. (1998). Odorants of virgin olive oils with different flavour profiles. *Journal of Agricultural and Food Chemistry*, 46(7), 2754–2763.

Ridolfi, M., Terenziani, S., Patumi, M. and Fontanazza, G. (2002). Characterization of the lipoxygenases in some olive cultivars and determination of their role in volatile compounds formation. *Journal of Agricultural and Food Chemistry*, 50(4): 835–839.

Romero, I., García-González, D. L., Aparicio-Ruiz, R. and Morales, M. T. (2015). Validation of SPME–GCMS method for the analysis of virgin olive oil volatiles responsible for sensory defects. *Talanta*, 134: 394–401.

Salas, J. (2004). Characterization of alcohol acyltransferase from olive fruit. *Journal of Agricultural and Food Chemistry*, 52: 3155- 3158.

Salas, J. J. and Sanchez, J. (1998). Alcohol dehydrogenases from olive (Olea europaea) fruit. *Phytochemistry*, 48(1): 35–40.

Sanchez, J. and Salas, J. J. (2000). Biogenesis of olive oil aroma. In J. Harwood and R. Aparicio (Eds.), *Handbook of olive oil: Analysis and properties.* Gaithersburg, Maryland, USA: Aspen publications, Inc.

Santosa, M. (2010). *Analysis of sensory and non-sensory factors mitigating consumer behavior: A case study with extra virgin olive oil.* PhD Dissertation, Food Science: University of California, Davis.

Stefani, G., Romano, D., and Cavicchi, A. (2006). Consumer expectations, liking and willingness to pay for specialty foods: Do sensory characteristics tell the whole story?. *Food Quality and Preference*, 17(1–2): 53–62.

Tressl, R. and Drawert, F. (1973). Biogenesis of banana volatiles. *Journal of Agricultural and Food Chemistry*, 21 (4): 560–565.

Tsimidou, M. (1998). Polyphenols and quality of virgin olive oil in retrospect. *Italian Journal of Food Science*, 10: 99-116.

Vichi, S., Pizzale, L., Conte, L. S., Buxaderas, S. and Lopez-Tamames, E. (2003). Solid-phase microextraction in the analysis of virgin olive oil volatile fraction: modifications induced by oxidation and suitable markers of oxidative status. *Journal of Agricultural and Food Chemistry*, 51: 6564–6571.

INDEX

I

K

L

M

N

O

T

V

W

X

β